A Coast for All Seasons

A Naturalist's Guide to the Coast of South Carolina

Miles O. Hayes and Jacqueline Michel

Joseph M. Holmes

Illustrator

PANDION BOOKS

a Division of Research Planning, Inc.
Columbia, South Carolina

PANDION BOOKS

a division of Research Planning, Inc.

P.O. Box 328

Columbia, South Carolina USA 29202

Email: permissions@researchplanning.com

A CIP catalog record for this book has been applied for
from the Library of Congress.

ISBN 978-0-9816618-0-3

All photographs by Miles O. Hayes and Jacqueline Michel unless otherwise indicated.
Illustration and design by Joseph M. Holmes.

The paper used in this book meets the minimum requirements
of the American National Standard for Information
Sciences – Permanence of Paper for
Printed Library Materials,
ANSI/NISO Z39.48-1992

Printed in China by Everbest through
Four Colour Imports, Ltd., Louisville, KY

Front Cover: Northern end of Hunting Island, South Carolina at low tide on 30 January 1975.

Back Cover: Vegetated coastal dunes on northern end of Kiawah Island, South Carolina in October of 1974.

Ode to the Geologists

...and some rin up hill and down dale,
knapping the chucky stanes to pieces wi' hammers,
like sae mony road-makers run daft-
they say it is to see how the warld was made !

Sir Walter Scott

DEDICATION

Because coastal research is, by its very nature, a team effort, this book is dedicated to the group of graduate and undergraduate students who did most of the work (Larry, Duncan, Tom, et al.), the trench diggers (Colin, Todd, Dallon, et al.), the boat drivers (Travis, Cookie, Anton, et al.), the vibra-core teams (Tim, Bo, Jerry, et al.), the airplane pilots (Craig, Roger, Clarence, et al.), the course coordinators (Aileen, Patrice, Leslie, et al.), the magic mud dancers (Mike, Doug, Freeman, et al.), and the dozens of others who played the role of COASTAL HEROS over the years.

TABLE OF CONTENTS

Page

PREFACE ix

ACKNOWLEDGEMENTS..xi

1. INTRODUCTION 1

SECTION I: COASTAL PROCESSES AND LAND FORMS

2. ORIGIN OF THE COAST 7

3. COMPARISON WITH OTHER COASTLINES 19

4. COASTAL PROCESSES 25
    Tides..25
    Waves 27
    Hurricanes..38
    Earthquakes 42

5. MAJOR LANDFORMS OF THE COAST 43
    The Georgia Bight 43
    Barrier Islands...44
        Introduction...44
        Types of Barrier Islands...44
        Origin of Barrier Islands 46
    Tidal Inlets 50
    Estuaries 56
        Definition 56
        Characteristics..58
        Estuaries in South Carolina 61
    Deltas 61
    Sand Beaches...63
        A River of Sand..63
        The Beach Profile 63
        The Beach Cycle 67
        Physical Sedimentary Features and Biological Components 68
    Coastal Dunes..76

6. BEACH EROSION..81

Introduction 81
Major Causes of Beach Erosion 82
    Introduction 82
    Deficits in Sand Supply 82
    Sea-level Rise 87
Methods Commonly Used to Prevent Beach Erosion 91
Hurricane *Hugo* (1989) 94

7. ESTUARINE AND BACK-BARRIER HABITATS 103
General Morphological Model 103
  Marshes 103
    Introduction 103
    Freshwater Marshes 104
    Brackish Marshes 106
    Salt Marshes 106
  Tidal Flats 108
    Definition 108
    Exposed Sandy Tidal Flats 108
    Sheltered Mud Flats 109
    Dominant Biogenic Features 113
  Tidal Channels 114

8. THE FRESHWATER RIVER SYSTEMS 123
Description 123
Bottomland Hardwoods 125

9. CONTINTENTAL SHELF 129

10. POTENTIAL RECREATIONAL ACTIVITIES 131
Canoeing and Kayaking 131
Birding 131
Fishing 131
Shell Collection 132
Relaxing at the Beach 132

SECTION II: MAJOR COMPARTMENTS

11. THE GRAND STRAND 135
Introduction 135
Coastal Geology 135
The Great Beach Nourishment Debate 137
The Good 138
The Bad 139
The Ugly 140
To Do Or Not To Do 140
The Northern Border 142
Myrtle Beach State Park 142
Lewis Ocean Bay Heritage Preserve 145
Huntington Beach State Park 149
Brookgreen Gardens 150
Pawleys Island 152

12. THE SANTEE/PEE DEE DELTA REGION 155
Introduction 155

The Winyah Bay Estuary Complex 156
    Description 156
    Places to Visit 157
The Santee Delta 159
    General Morphology and Sedimentation Patterns 159
    Impact of Man 162
    Places to Visit 165
Cape Romain 170
    Morphology and Origin 170
    Places to Visit 175
Bulls Bay 176

13. THE BARRIER ISLANDS 179
Introduction 179
Bull Island Area 180
    Morphology and Origin 180
    Places to Visit 182
Isle of Palms/Sullivans Island 186
    Morphology 186
    History 188
    Places to Visit 188
Folly Beach/Morris Island Area 191
    Morphology, Origin, and History 191
    Places to Visit 196
Kiawah and Seabrook Islands 200
Kiawah Island 200
    History 200
    Geomorphology and Erosional/Depositional History 201
    The Beach 204
    Tidal Inlets 207
    Salt Marsh and Tidal Channels 207
    Stratigraphy 207
    Places to Visit 209
Seabrook Island 211
    Discussion 211
    Places to Visit 216
Edisto Beach Area 217
    Introduction 217
    Edisto Beach Recurved Spit 217
    Landward-Migrating Barrier Islands to the Northeast 219
    Places to Visit 220

14. THE LOW COUNTRY 225
Introduction 225
The ACE Basin 226
    Morphology and Origin 226
    Places to Visit 229
The Hunting/Fripp Islands Headland 235
    Morphology, Origin, and History 235
    Places to Visit 238
Port Royal Sound 245

Morphology and History 245
    Places to Visit...247
Hilton Head Headland 248
    Morphology and History 248
    Places to Visit...250
Savannah River Delta...254
    Morphology 254
    Places to Visit...255

15. THE RIVERS AND SWAMPS 259
    Introduction...259
    Places to Visit 259

16. FUTURE OF SOUTH CAROLINA'S BARRIER ISLANDS 265

RECOMMENDED ADDITIONAL READING...267

REFERENCES CITED..271

INDEX 281

ABOUT THE AUTHORS..285

PREFACE

In the mid-1960s, while I was a graduate student in the Geology Department of the University of Texas (UT), I had a part-time job in the Defense Research Laboratory at UT which allowed me to conduct a study of the coastlines (and inner continental shelves) of the world as part of their classified anti-submarine warfare program. As a result of that work, I began to distinguish among coastlines like those in Texas (low-relief coastal plains and deltas, which we eventually called depositional coasts) on the basis of tidal range (vertical distance between high and low tide). I recognized that depositional coasts with a large tidal range were distinctly different from those with a small tidal range. An outside advisor of my dissertation committee, Dr. W. Armstrong Price, had pioneered these ideas, but I then had a statistical basis for extending them.

As I was completing my Ph.D. study of the south Texas coast, which has a small tidal range, my ambition was to eventually carry out similar studies of coastlines around the world with intermediate and large tidal ranges. Therefore, the very first research grant I applied for, with the U.S. Navy research program, was to study the depositional coastline of South Carolina with its intermediate-range tides. That research proposal was turned down flat. Also, there were no job openings requiring my skills in any of the geology departments of the universities in the southeastern United States.

Fortunately, shortly thereafter I was able to secure a job at the University of Massachusetts. The relatively unstudied coastline of New England, which has a tidal range similar to South Carolina, was at the disposal of me and my versatile crew of graduate students. It was a great experience and we learned a lot, but the South Carolina coast still beckoned. Therefore, in the fall of 1972, a group of my students and myself migrated to Columbia, South Carolina to join the Geology Department of the University of South Carolina (USC). During the ensuing years at USC, my students produced over 25 M.S. and Ph.D. theses on studies of the South Carolina coast.

Fortunately again, a couple of years after I arrived in Columbia, Jacqui Michel joined our research trips from time to time while completing her Ph.D. in geochemistry at USC under Dr. Billy Moore. Jacqui and I were married in 1977, eventually forming a science technology company called Research Planning, Inc. (RPI), and have continued to work together on the South Carolina coast ever since. Also, during the decades of the 1980s and 1990s, we were able to fulfill my original ambition to study shorelines with different tidal ranges, culminating with studies of shorelines with tidal ranges >30 feet in Chile, Alaska, and France.

A happy sideline to all this was a contract through RPI with the American Association of Petroleum Geologists (AAPG) and Schlumberger Well Services to conduct week-long field seminars along the South Carolina coast, sometimes as many as 6 or 7 a year. Therefore, these seminars, which started in September, 1976 and continued for 30 years (last one in the fall of 2005), were always a welcome respite from the travels in numerous countries in the Middle East and Africa, from airplane crashes in Alaska, and so on. It was always most pleasant and relaxing to be back riding the boat along the tidal channels and walking the beaches of the South Carolina coast. The discussions we had with the several thousand geologists from all around the world (with a majority from Canada) who attended these seminars were always very exciting, rewarding and thought provoking. Guess what? We talked about the effect of the tides a lot!

Whereas we won't be taking you on any boat rides, beach walks, or overflights in a small airplane, which was always a part of our field seminars, we hope the descriptions and illustrations in this book will convey a little of the excitement that we always feel when we are out in the field on our beautiful coast for all seasons.

Miles O. Hayes, Columbia, S.C., 20 January 2008.

ACKNOWLEDGEMENTS

A special thanks to Tim Kana and Walter J. Sexton for their reviews of an earlier draft of this book. Also, they spent many days, weeks, and months with us both in the field on the South Carolina coast and in our business endeavors at RPI. They are also acknowledged for their scientific contributions over the years that have added so much to our understanding of how the coast was made.

Much of the information presented here is based upon research funded by a number of sources, including the U.S. Army Corps of Engineers, U.S. Army Research Office, Office of Naval Research, National Science Foundation, South Carolina Sea Grant Consortium, South Carolina Coastal Council, National Oceanic and Atmospheric Administration (NOAA), Geology Department of the University of South Carolina, RPI International, Inc., and Research Planning, Inc. The continuing support of the American Association of Petroleum Geologists and Schlumberger Well Services during the 30 years we taught geological training seminars on the coast is also gratefully acknowledged. In addition to Tim Kana, the logistics coordinators for these courses over the years included Aileen Duc, Patrice Cunningham, and Leslie Sauter. The expertise of Mike Bise, who managed and executed the field logistics of the seminars for so many years, rescued us from mishaps time and time again.

Associates in the Coastal Research Division of the Geology Department of the University of South Carolina and at RPI, who carried out much of the research on the coast discussed in this book, included (in addition to Kana and Sexton) - Pulak Ray, Mike Stephen, Dag Nummedal, Larry Ward, Gary Zarillo, Bob Finley, Duncan FitzGerald, Dennis Hubbard, John Barwis, Tom Moslow, Aileen Duc, Chris Ruby, Stan Humphries, Ray Levey, Cary Fico, Jeff Knoth, Craig Shipp, Pete Reinhart, Bob Tye, Martha Griffin, John Hodge, Dave Nelligan, Mike Svetlichney, Chris Reel, Tom Freeman, Cathy McCreech, C. Y. McCants, Dan Domeracki, Helen Mary Johnson, Doug Imperato, Steve Wilson, Leita Jean Hulmes, Tim Eckard, Todd Montello, Colin Plank, and Dallon Weathers.

The Pandion Books Division of RPI provided funding, as well as staff support (notably Joe Holmes, Wendy Early, and Jack Moore), for the final publication of this book.

1 INTRODUCTION

Our goal for this book is to introduce you to the natural setting of the shoreline of South Carolina, one of the most beautiful in all the country and one you can enjoy visiting during all seasons of the year (see satellite image of the entire coastline in Figure 1). Michel was raised on the shores of Charleston Harbor before gaining her Ph.D. in Geology at the

The overall approach and underlying theme of this book are based on a description of the **physical characteristics** of the coast, because that physical framework is the basic skeleton upon which other relevant features, such as coastal wetlands and tidal flats, rest. For example, certain fundamental questions, such as why do some barrier islands

---

There are two ways you may want to consider using this book:

1) If you wish to gain an introductory understanding of the coastal processes and land forms, you should spend some time right away reading Section I.

2) On the other hand, if you are visiting the coast and you want to learn about a specific area, you can look up that area in Section II. Then, if you have questions about some aspect of that particular area, such as why the beach is eroding, you can read up on those topics in Section I.

University of South Carolina (while participating in many research projects on the South Carolina coast), and Hayes made his first appearance at the coast thirty five years ago (1972) as the Director of the Coastal Research Division of the Geology Department of the University of South Carolina. Therefore, we have spent many decades trying to decipher the mysteries of this majestic coast for all seasons. Hopefully, we have been able to adequately show in this text the enthusiasm for and understanding of this coastline that it so richly deserves.

erode while others do not, cannot be answered without a firm understanding of the physical processes that formed the islands. However, none of the major coastal habitats, including their biological components, are neglected as we explore the coastline from the North Carolina border to the Savannah River.

Before getting further into this discussion, we need to introduce the term **geomorphology** (*geo* = earth; *morph* = shape, form; *ology* = to study), a scientific discipline devoted to understanding the origin and three-dimensional shape of the landforms of the earth. The discipline of **coastal geomorphology** is focused on the understanding of how the different coastlines of the world originated, as well as their resulting forms and organization. Both of the authors of this book are coastal geomorphologists, but our interests and professional activities have provided for us a broad experience in other aspects of **coastal ecology**. We are ardent birders and the work of our company, Research Planning, Inc. (RPI) of Columbia, South Carolina, in areas such as beach erosion, oil-spill response, and natural resources mapping, has broadened our understanding of coastal ecosystems.

The book begins in Section I with a general discussion of the origin of the South Carolina shoreline, starting as far back as over 300 million years ago when the rocks that presently underlie the coastal areas at depth were part of the supercontinent **Pangaea**. This history of the evolution of the coast ends with a discussion of the factors that have shaped the modern coastline since the last rise in sea level that resulted from the melting of the glaciers and ice fields of the last **Ice Age**. Next, this coastline is compared with others around the world, identifying the key factors that give the South Carolina shoreline its distinguishing characteristics.

The rest of Section I examines the dynamic physical processes that shape the present shoreline, namely waves and tides, plus an introduction to the influence of hurricanes. Next, the major landforms of the South Carolina coast are described, specifically – barrier islands, tidal inlets, estuaries, and deltas. South Carolina has an abundance of sand beaches, thus there is a detailed discussion of the beach cycle and a review of the erosional impacts of **hurricane *Hugo* (1989)** on the South Carolina beaches. Specific habitats of the major landforms, such as coastal dunes, marshes, tidal flats, and tidal channels, are also treated in some detail. Section I concludes with a short discussion on some of the recreational options in the area.

In Section II, we discuss and identify oppor-tunities for naturalists to learn about and enjoy the diverse ecosystems that exist within the **four major geomorphological compartments** of the coastal zone. The description of each compartment begins with notes on the morphology and environments of the compartment, as well as some information on its human history where relevant. A list and discussion of the **key places to visit** conclude the coverage of each of the four compartments, which are labeled on Figure 1 and introduced below:

- The **Grand Strand** – This arcuate shoreline with nearly continuous sand beaches extends from the North Carolina border to the southern edge of Pawleys Island. Included in the compartment are two excellent state parks with expansive sand beaches, **Myrtle Beach** and **Huntington Beach State Parks**. Among others, two of the major topics discussed in this section include the **value of beach nourishment projects** and the origin of **Carolina Bays**.

- The **Santee/Pee Delta Region** – The largest river delta on the east coast of the United States creates a bulge in the shoreline that extends from the south end of Pawleys Island to the southern margin of Bulls Bay. Discussion topics covered in this section include historic rice farming practices, detailed characteristics of the delta, and the origin of the major North Carolina/South Carolina **Capes** (Hatteras, Lookout, Fear, and Romain).

- The **Barrier Islands** – This chain of 14 barrier islands extends from Bull Island in the north to the south end of a recurved spit at Edisto Beach. Discussion topics covered in this section include the details on the origin of Kiawah Island and the impact of the Charleston Harbor jetties on shorelines to the south, particularly at Folly Beach and Morris Island. Two visitor-friendly county parks, **Folly Beach Park** on Folly Island and **Beachwalker Park** on Kiawah Island, have inviting, wide and flat sand beaches.

- The **Low Country** – This complex shoreline

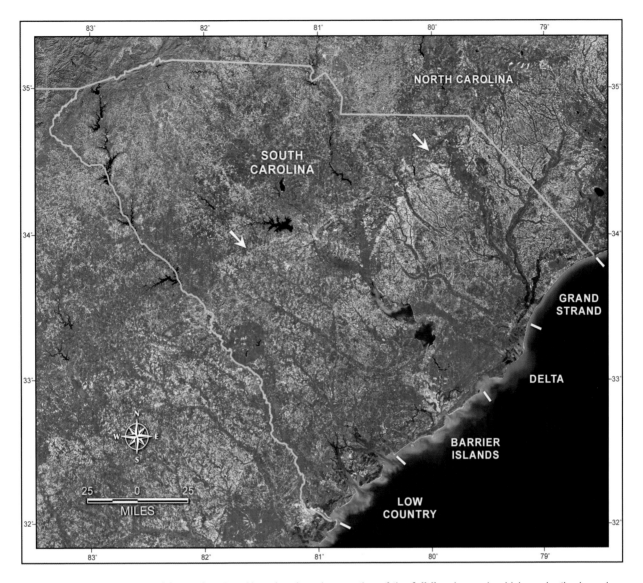

FIGURE 1. Satellite image of South Carolina. Note the clear demarcation of the **fall line** (arrows), which marks the boundary between the Piedmont and Coastal Plain Physiographic Provinces. The **swamps** and **upper bottomland hardwoods** on the flood plains of the rivers that cross the coastal plain are also clearly visible as solid red color on this infrared image. The boundaries of the four major morphological compartments of the coast (Grand Strand, Santee/Pee Dee Delta Region, Barrier Islands, and Low Country), which are discussed in detail in the text, are also shown. Landsat image mosaic acquired 1999-2001, courtesy of U.S. Geological Survey.

consists of two major estuarine complexes, two protruding headlands, and a large delta system that extends from the northern entrance of St. Helena Sound to the Savannah River. Three national wildlife refuges and a large state park, **Hunting Island State Park,** offer outstanding opportunities for birding, as well as opportunities to view extensive salt marshes and tidal flats. Two very famous barrier islands,

Hilton Head and Daufuskie, are also located in this compartment.

Section II concludes with a discussion of several of the major rivers and swamps within the coastal plain, as well as some thoughts on the possible future of the barrier islands along the coast.

SECTION I
Coastal Processes and Land Forms

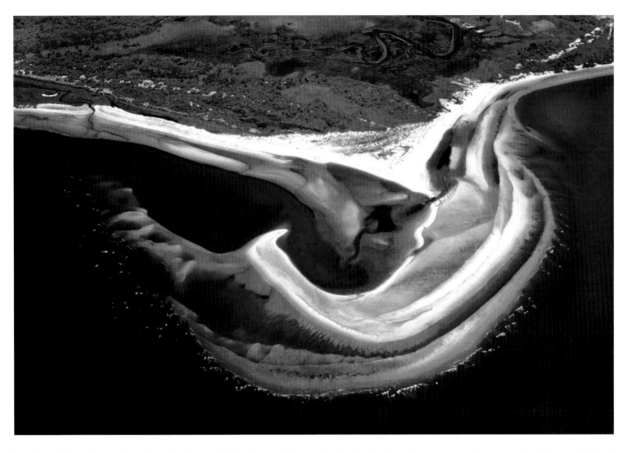

Low-tide infrared photograph of a large intertidal sand bar welding to the beach on the northeast end of Kiawah Island (taken on 29 May 1980).

2 ORIGIN OF THE COAST

Once upon a time at the spot where the authors of this book have a mountain house in western North Carolina, in place of the now serene Blue Ridge Mountains, stood a stark, snow-capped mountain chain as high as the modern Himalayas. Eventually, that high mountain range was eroded, and, in the process, provided the bulk of the sediment that now comprises the Coastal Plain of South Carolina, including the light gray sand on the beaches. Unraveling the long history of how the present coastline originated has fascinated geologists for many decades. The story is rather complicated, and there are still many unanswered questions regarding details, such as the exact age and timing of the different geological episodes. However, the following brief listing of this chain of events, based on a variety of sources (sciencedaily. com; earthquakes.usgs.gov; news.softpedia.com; sciencedaily.com; geology.er.usgs.gov; Glen, 1975; and Hatcher, 2005), provides a general introduction to the subject. See also the time line in Figure 2.

There is still some question about the exact timing of all this, but conventional wisdom has it that about 400 million years ago, there were two main land masses separated by the Rheic Ocean. In the south sat **Gondwana**, a supercontinent consisting of South America, Africa, India, Australia, and Antarctica, and in the north sat **Laurasia**, made up of North America, Greenland, Europe, and part of Asia. There is still some dispute about the exact makeup of these two landmasses.

[NOTE: In the winter of 1964/65, co-author Hayes led a geological expedition in Antarctica to examine a sequence of sedimentary rocks called the Beacon Sandstone of Permian age (around 250 million years old). These rocks were very similar to other Permian units in South Africa and India, containing huge petrified trees that grew in tropical forests, among other artifacts. The correlation of these very similar sandstones by the geologists at that time was supporting evidence for the original makeup of the supercontinent Gondwana. Those studies also documented the eventual wandering of the fragment of that original continent now known as the separate continent of Antarctica from warmer climes near the equator to the vicinity of the present south pole.]

Around 300 million years ago, during the Carboniferous Period of the Paleozoic Era (see geological time scale in Table 1), these two landmasses collided to form a single large continent called **Pangaea** (see Figure 3A). In the process, the Appalachian Mountains were formed (Dietz and Holden, 1970). These mountains are said by some to have been as high as the Himalayas and by others at least as high as the modern Rocky Mountains. During this time interval, the ancient igneous and metamorphic rocks that presently underlie the

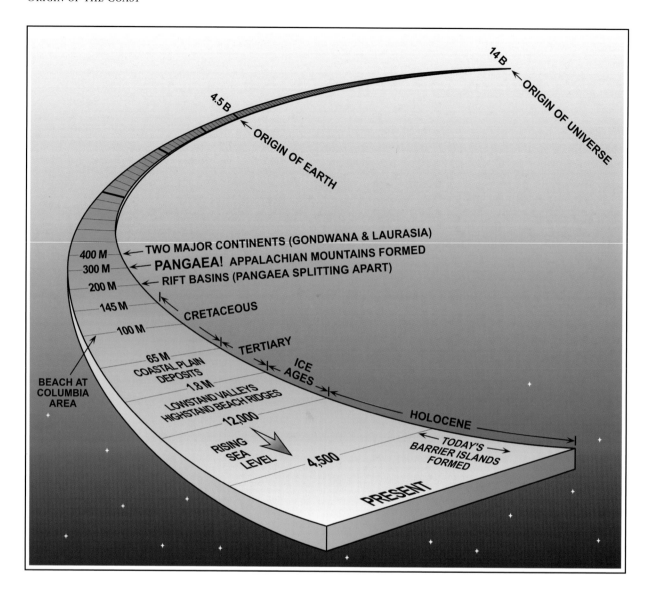

FIGURE 2. Time line for origin of the coast of South Carolina. See Table 1 for nomenclature and timing of the geological events.

sediments and sedimentary rocks of the South Carolina Coastal Plain were intensely folded and faulted.

[NOTE: The three major classes of rocks – igneous, sedimentary, and metamorphic – were first defined by the pioneering Scottish geologist James Hutton in the mid 1700s. He recognized a class of rocks called **igneous rocks** that crystallized from an extremely hot molten mass of material (magma) that was formed as if by fire (*ignis* = fire). One class of igneous rocks, which includes granite, develops by slow cooling of the molten mass at great depths

within the earth's crust. As a result of the slow cooling, large crystals of individual minerals form that can be as much as an inch or so in diameter. These minerals, typically including feldspar and quartz, eventually coalesce to form a rock mass, which may later be pushed up to the earth's surface by mountain building processes. A second class of igneous rocks, called volcanic rocks (e.g., basalt and rhyolite), crystallize rapidly as a result of molten lava being extruded suddenly onto the earth's surface. The crystals of these minerals that form so rapidly are, for the most part, too small to be seen

TABLE 1. Geologic time. Time intervals are listed in millions of years before the present (from numerous sources).

Era/Period/Epoch			Time (Million yr ago)
Cenozoic era "Recent Life"	Quarternary period	Holocene epoch	0.012-0
		Pleistocene epoch	1.8-0.012
	Tertiary period	Pliocene epoch	6-1.8
		Miocene epoch	26-6
		Oligocene epoch	38-26
		Eocene epoch	55-38
		Paleocene epoch	65-55
Mesozoic era	Cretaceous period		145-65
	Jurassic period		205-145
	Triassic period		250-205
Paleozoic era	Permian period		290-250
	Carboniferous (Mississippian/Pennsylvanian) period		355-290
	Devonian period		410-355
	Silurian period		438-410
	Ordovician period		505-438
	Cambrian period		545-505
Precambrian	Proterozoic era		1500-545
	Archaeozoic (Archean) era		4500-1500

by the naked eye. On the other hand, **sedimentary rocks** are most commonly formed by weathering and erosion of preexisting rocks on the earth's surface. This process creates sediment that can be transported by water, or wind in some cases, to be deposited in large masses on river deltas, beaches, and so on. Once deposited, these most commonly sand-sized sediments may become buried where they are consolidated by chemical cementation and other processes. The most abundant sedimentary rocks formed by this process are called sandstones. Limestones, which are also sedimentary rocks, usually develop in marine waters by a combination of chemical precipitation and accumulation of the hard parts of some marine organisms, such as sea shells. The third class, **metamorphic rocks,** results from dramatic changes in igneous and sedimentary rocks affected by heat, pressure, and water that usually results in a more compact and more crystalline condition. These changes usually take place at significant depths below the earth's

surface. Examples of metamorphic rocks include: 1) slate, derived from a sedimentary rock called shale (originally composed of silt and clay); 2) marble, derived from limestones; and 3) gneiss, a banded, micaceous rock typically derived from granite.]

Between 225-190 million years ago, during the Triassic Period, the major land mass of Pangaea started to split apart along the old collision zones. This pulling apart created some **rift basins** in parts of the area now known as the eastern United States similar to the ones presently occurring in south-central Africa. These rift valleys filled with sediments eroded from the uplifted blocks along the rift margins. Some of the most noteworthy of these Triassic Rift Basins occur in Massachusetts and North Carolina. However, the map in Figure 4 shows the presence of one of these rifts along the South Carolina/Georgia border (the South Georgia Rift). The sedimentary rocks in the South Georgia Rift are the oldest sedimentary rocks that have not been metamorphosed (i.e., changed dramatically

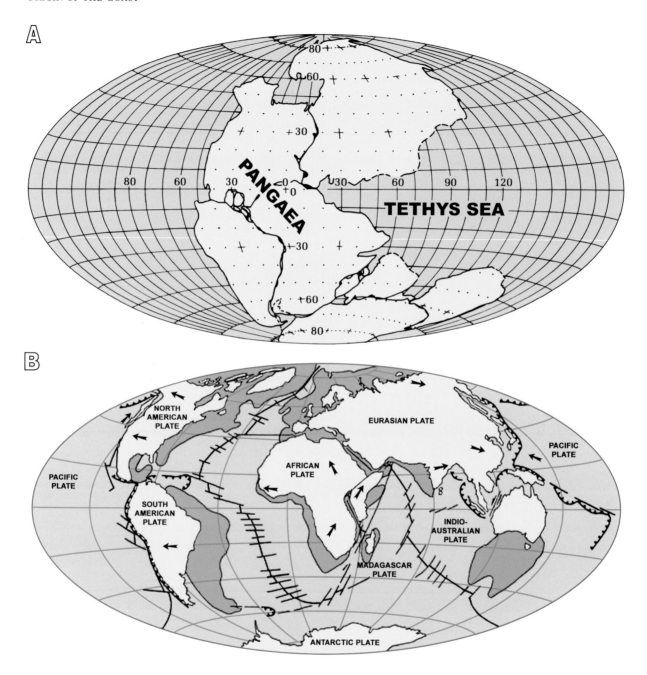

FIGURE 3. Two of the basic elements in the geological history of South Carolina as we now know it. (A) Outline of all the continents that assembled to form **Pangaea**, a universal landmass as it existed at the end of the Permian Period (Table 1) around 250 million years ago [as defined by Dietz and Holden, (1970); modified after Glen (1975: Figure 8.8)]. (B) The major continents in their present position. Arrows indicate the directions the different land masses have moved since they first started "drifting apart" around 200 million years ago. The blue "shadows" represent previous locations of the continents as they moved along.

by high heat and pressure) in the state of South Carolina.

Around 100 million years or so ago as the continents continued to drift apart (see Figure 3B, which shows the final product of this drifting), a huge arm of the world ocean, called the **Cretaceous Seaway,** covered much of what is now North America, extending north to south through the

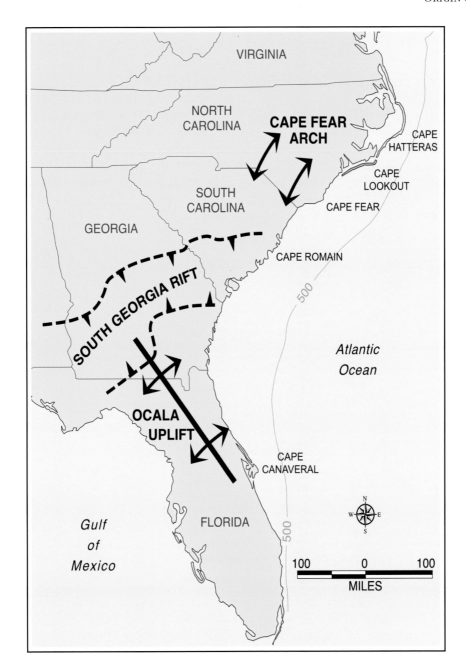

FIGURE 4. Coastline of the southeastern United States, illustrating the major bend in the shoreline known as the Georgia Bight. Some of the principal tectonic elements, the Cape Fear Arch to the north and the Ocala Uplift to the south, as well as the South Georgia Rift (of Triassic age; Table 1), are shown. Tectonic elements delineated by LeGrand (1961).

middle of the continent. All of the present United States south of the southern edge of the Appalachian Mountains was flooded by this large sea, which extended to the east into the ocean being created by the continents drifting apart. As shown in Figure 5, during the Late Cretaceous, the shoreline was near where Columbia, South Carolina is now located. The

Cretaceous sedimentary rocks in South Carolina were deposited in this ancient sea, containing fossils of oysters and other marine animals, as well as dinosaur teeth and bones washed from the shore. These sediments can be seen today on the surface in a relatively narrow band at the upper edge of the coastal plain, and they achieve thicknesses of

11

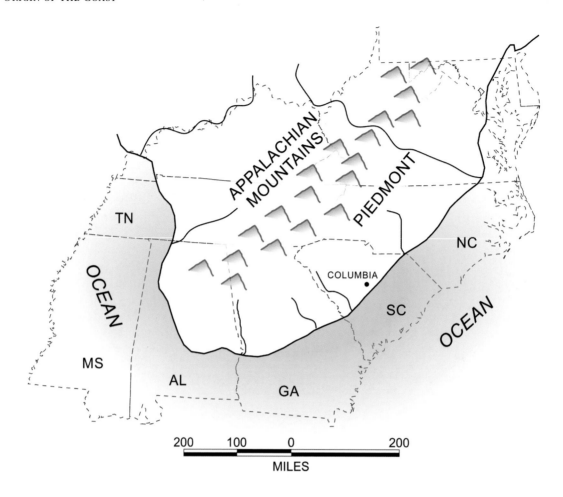

FIGURE 5. Approximate location of the Cretaceous Seaway in the eastern United States around one hundred million years ago. The shoreline of the Seaway was located near the present town of Columbia, South Carolina. Modified after D. T. King, Jr.

several hundred feet at depth in the lower coastal plain (Figure 6). These sedimentary rocks are important ground-water aquifers throughout the coastal plain.

During the **Tertiary Period** (between 65 and 1.8 million years ago) the **Coastal Plain** of the region we know as South Carolina was transgressed (flooded) and exposed several times. That span of time is described succinctly on the paleoportal. org web site as follows: *Erosion of the Appalachian Mountains supplied much of the material that makes up the Tertiary rocks of South Carolina. Fossils from these rocks indicate that the climate of the Early Tertiary (Paleocene and Early Eocene) was warm and tropical. Worldwide cooling events generated a more moderate climate for South Carolina during the Late Eocene. But by the Middle Tertiary (Oligocene), the climate began to warm again, as evidenced by fossils of whales and large crocodiles. Various marine and terrestrial fossils show that the climate was tropical to subtropical, much like modern Florida, by the Late Tertiary (Pliocene).* Tertiary rocks, which are predominately sandstones and shales, but do contain some significant limestone deposits, are exposed at the surface over much of the Middle and Upper Coastal Plain. They also achieve thicknesses of several hundred feet at depth in the Lower Coastal Plain (Figure 6). Depositional environments ranged from deltas and other shoreline deposits to offshore limestones.

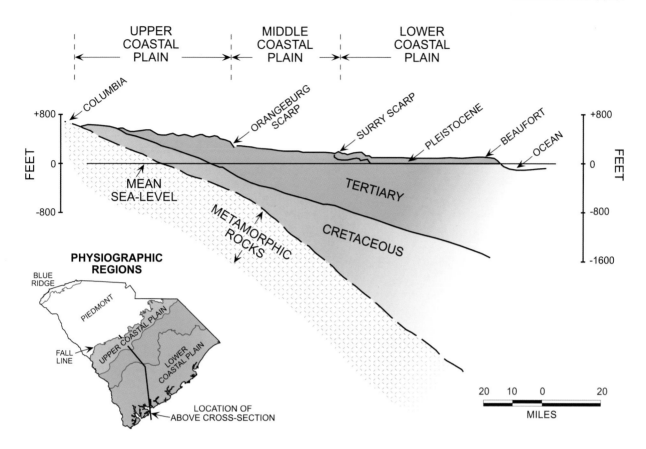

FIGURE 6. General cross-section of the sedimentary rocks and sediments that underlie the Coastal Plain Physiographic Province of South Carolina. Location of the three physiographic provinces in the state is shown in the lower left. The Lower Coastal Plain is characterized by a relatively flat surface underlain by Pleistocene sediments. The Middle and Upper Coastal Plain, which are underlain by surficial outcrops of Tertiary and Cretaceous rocks and sediments, show considerably more topographic relief than the Lower Coastal Plain. These older sedimentary rocks abut against the igneous and metamorphic rocks of the Piedmont Province along the fall line (see Figure 1). Highly modified after Colquhoun et al. (1983).

A major change in climate that initiated the beginning of the Ice Ages (the **Pleistocene Epoch**) occurred around 1.8 million years ago. The Pleistocene Epoch ended around 10-16,000 years ago, when the last of the major glaciations ceased. Within the Pleistocene Epoch, four major glacial events occurred, during which time ice covered large areas of the North American continent (as far south as the Ohio River in the eastern United States). The approximate median peak times for these glacial events, expressed in years before the present (B.P.), were: 650,000 for the Nebraskan glaciation; 450,000 for the Kansan; 150,000 for the Illinois; and 60,000 for the Wisconsin. During each of these glaciations, sea level dropped to significantly lower levels than it is today; during the last glaciation period it was 350 feet lower! When sea level was lower, during what is referred to as **lowstands**, the rivers along the east coast carved deep valleys across the coastal plain and out onto the present continental shelf. The location of these valleys on the coastal plain of the Georgia Bight is shown in Figure 7. The glaciations were separated by warming periods, called interglacials, when sea level rose. During these warming periods, the highest sea levels were higher than it is today, with each succeeding interglacial having lower and lower **highstands**. Sea level during the last interglacial, the Sangamon (peak around 120,000 years B.P.), stood around 10 to 12 feet higher than it is today.

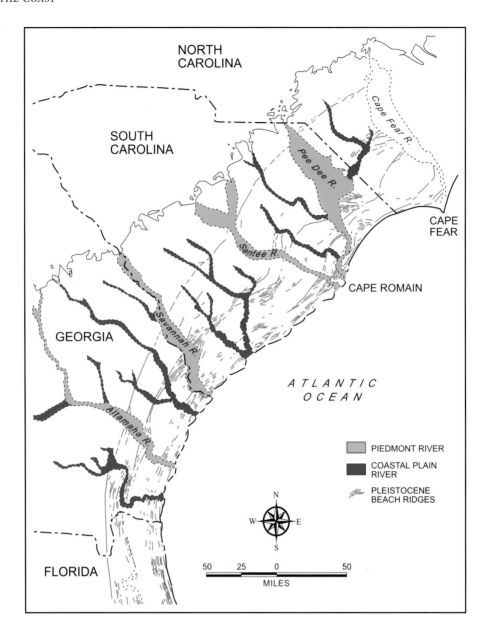

FIGURE 7. Lowstand valleys of the coastal plains of South Carolina and Georgia. The valleys were carved when sea level was lowered during the major glaciations of the Pleistocene Epoch. The bulk of the flood plains that now occupy these valleys were formed during the last major rise in sea level, which started about 12,000 years ago. The valleys of the four piedmont, or red-water, rivers (Pee Dee, Santee, Savannah, and Altamaha), are shown in orange, and those of the major coastal plain, or black-water, rivers are shown in blue. The red lines indicate the position of beach ridges or erosional scarps formed when highstands of sea level occurred between the major glaciations. Highly modified after Winker and Howard (1977).

During these highstands, waves eroded scarps and deposited linear beach ridges inland of the present shoreline. The old beach ridges, shown in red on Figure 7, extend inland as far as 40 miles. Note the zones of parallel beach ridges several miles wide formed during the Sangamon interglacial, that are located only a short distance landward of much of the present shoreline. Pleistocene sediments blanket the Lower Coastal Plain (Figure 6), which is very flat. During some of the lowstand episodes, the actual shoreline was near or at the edge of the present continental shelf, a distance of 60-70 miles from the present shore. Also, during episodes when sea level was relatively constant for a while during its covering and uncovering of the continental shelf, coastal features such as river deltas and barrier-island complexes were deposited; remnants of them are still present on the continental shelf. Such events have created a very complex topography of the shelf (see Figure 8).

The major ice-melting episode that started about 10-16,000 years ago marked the beginning of the **Holocene Epoch** at which time sea level was approximately 350 feet below its present position. A considerable amount of research on this topic is still underway, and the exact dates involved are subject to change as more data are produced. Without any doubt, however, as the melting proceeded, sea level rose rapidly, as much as 2 feet/century, reaching near its present level in South Carolina around **4,500**

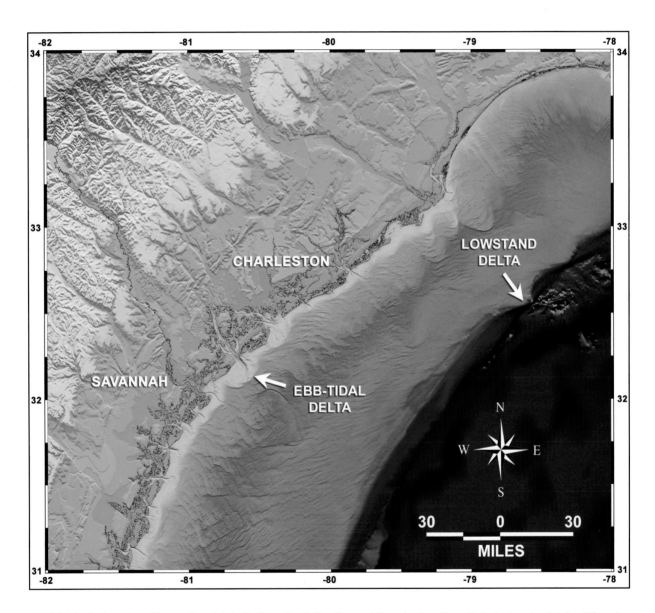

FIGURE 8. Bathymetry of the continental shelf off the South Carolina and Georgia shorelines. Note the lowstand delta at the edge of the continental shelf (arrow), as well as the numerous relict deltas further landward on the shelf, off the present mouths of the Santee and Pee Dee Rivers. The massive ebb-tidal delta off Port Royal Sound (arrow) is also clearly shown. From the National Geophysical Data Center (Coastal Relief Model).

B.P. (the exact date is somewhat in question). Since a near **stillstand** that occurred at that time (4,500 B.P.), the bulk of the major Holocene landforms on the coast (e.g., deltas and barrier islands) have formed. However, as shown in Figure 9, there have been several minor sea-level fluctuations, up to four feet or so, in the past 4,500 years. The remainder of this book mostly concerns features formed during that last, relatively **short time interval**.

This long evolution has left a final imprint on the state of South Carolina, which consists of the three major physiographic provinces shown in Figure 6. Just a sliver of the **Blue Ridge Province** occurs in the extreme northwest portion of the state. This

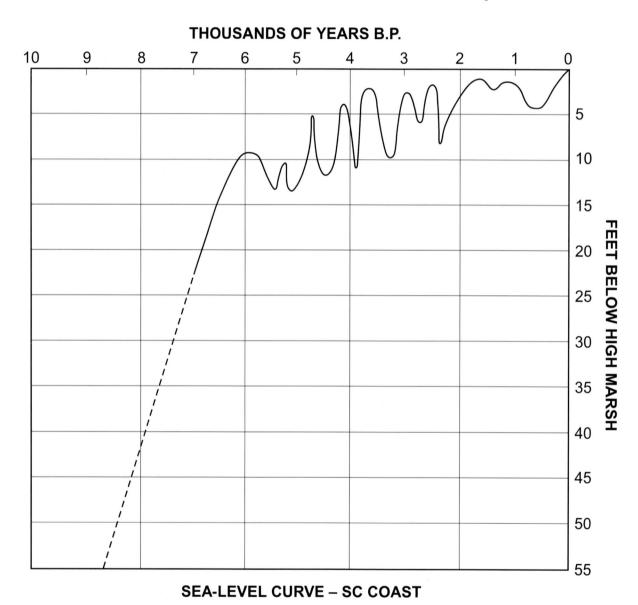

THOUSANDS OF YEARS B.P.

SEA-LEVEL CURVE – SC COAST

FIGURE 9. Changes in sea level along the South Carolina coast over the past 6,000 years, based on detailed, combined archeological and stratigraphical research (after Colquhoun and Brooks, 1986). The dashed extension of the curve is our approximation. A similar curve was derived for the Georgia coast by DePratter and Howard (1980), but DePratter (pers. comm.) indicates that this curve needs some upgrading. Therefore, some changes are probably in order. Nevertheless, some of the lows in the curve between 2,000 and 4,000 years before present have had a significant effect on the evolution of the barrier islands in the Georgia Bight.

mountainous region has elevations ranging from 1,000 feet in the foothills to 3,554 feet at Sassafras Mountain, the highest point in South Carolina. The Blue Ridge Mountains of South Carolina are but a small portion of the Appalachian Mountain system. The Blue Ridge Province is bordered on the south by the **Piedmont Province.** Piedmont is derived from a French word meaning "foot of the mountains." This Province consists of a 100-mile-wide belt between the Blue Ridge and Coastal Plain Provinces. The fault line called the Brevard Zone separates the Piedmont from the Blue Ridge. The rock types in the Piedmont are primarily metamorphic, mainly schists, gneisses, and slates, and some granite igneous rocks that intruded into cracks and joints in the existing rock about 200 million years ago. The **Coastal Plain Province**, which has already been discussed in some detail, covers more than half of the state. The Upper and Middle Coastal Plain has some relief with hills scattered here and there; but the Lower Coastal Plain, with its surface sediments of Pleistocene age, is nearly flat and featureless.

The **climate** of the three physiographic provinces is summarized briefly below:

- The whole state has a **humid subtropical climate**, with an average precipitation in the coastal plain of about 49 inches/year.
- There is more rainfall in the Blue Ridge Province, with up to 80 inches/year.
- Regional rainfall related to the passage of large fronts is common in the winter months, whereas rainfall in the summer is usually associated with thunderstorms.
- The average January temperatures range from 50 degrees along the coast to 38 degrees in the mountains. Average July temperatures are 81 degrees along the coast and 71 degrees in the mountains.

3 COMPARISON WITH OTHER COASTLINES

All coastlines have a rather complicated history, such as the one just described for the coast of South Carolina. To put the previous discussion in context, a brief look at how other coastlines around the world have formed and how the South Carolina coast compares with them follows. The authors of this book have been studying coastlines for many years, with Hayes' analysis of the coastlines of the world for the U.S. Navy in 1960-65 being a starting point. Since that time, the authors have worked on the shorelines of over 50 countries, including almost all of the shoreline of North America, the Middle East, parts of Central and South America, and West Africa. Hayes also supervised Masters and Doctoral research of 72 graduate students, most of which had to do with coastlines, while a Professor at the Universities of Massachusetts and South Carolina. Michel's specialty of oil-spill response has enabled her to carry out research in a wide variety of shoreline settings.

When we started our research in coastal geomorphology, understanding how the coast was made was almost virgin territory, at least in the United States. Hypotheses on how barrier islands evolve, how the tides and waves shape the coast, and how coastal storms change things were there for the alert observer to deduce, assuming enough examples had been seen and that the global processes had been properly accounted for. We were using what Comet (1996) referred to as "one of Aristotle's two forms of logical inference" – namely, inductive reasoning from the observation of a generalized pattern or distribution in order to develop a principle or law. In other words, we started with a large number of more or less random observations, not with a detailed data set on a specific topic. However, we did eventually try to figure out how to collect meaningful data sets to verify or disprove those deduced principles or laws.

Based on this experience, we conclude that the following hierarchy of factors determine the geomorphological makeup of any shoreline – in decreasing order of importance: 1) global tectonic crustal movements; 2) hydrodynamic regime; 3) climate; 4) sediment supply and sources; and 5) local geological history and sea-level change.

When Hayes was a graduate student at the University of Texas in the 1960s, very few of the professors in the elite, Ivy League universities "believed in" continental drift, although the idea had been seriously proposed and discussed by scientists since almost the beginning of the 20th century. The problem for the "non-believers" was that no credible mechanism had been proposed to explain why the continents had drifted apart. Now, of course, the concept of **plate tectonics** provides the basis for almost all current thinking relative to most topics in geology, including the mechanism for continental drift. Geophysical evidence, such as increasing age of parallel magnetic bands of volcanic rocks on the sea floor away from the mid-ocean ridges, provided indisputable evidence that the continents are indeed drifting apart. The exact mechanism causing this movement is more in doubt. The favored theory at this time is that huge individual slabs of rock (called **plates**, the existence of which is not in doubt) are slowly moved apart by convection cells of hot, softened mantle material below the rigid plates. That is one favored theory for why the plates move. There are other more complex ones. One fine day all of this will be figured out. Meanwhile, let us get back to the indisputable fact that the continents are moving.

Tectonics is defined as the forces involved in or producing deformation of the earth's crust, such as

folding or faulting of the rocks. The cross-section given in Figure 10 shows one of the basic concepts of plate tectonics, in which a plate made up of a continental crust (e.g., the North American plate, which is relatively light because of the abundance of aluminum, silica, sodium, etc. in the rocks) is riding over a sinking oceanic plate (e.g., the Pacific plate, which is relatively heavy because of the abundance of iron, magnesium, etc. in the rocks). If we look at one of the simpler examples, such as the South American plate, a striking contrast can be seen between the western shoreline of the continent and the eastern shoreline. Along the western shore, where the continental and oceanic plates collide, or leading edge of the continental plate, a series of young mountain ranges flank the coast. However, on the eastern shore, or the trailing edge of the

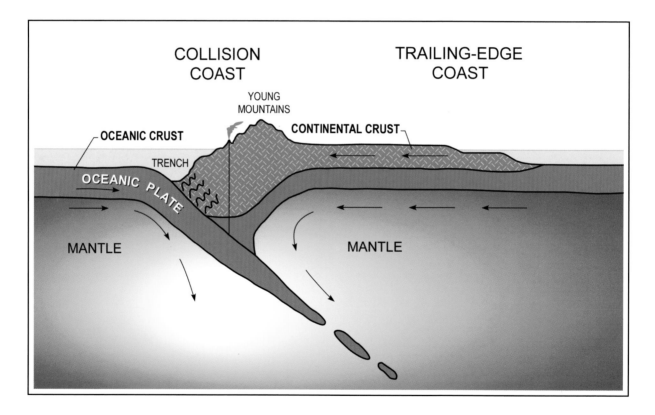

FIGURE 10. Cross-section illustrating the collision of continental and oceanic plates. At the collision point, the continental plate rides over the oceanic plate, creating a high, young mountain range. Shorelines along the collision juncture tend to be dominated by rocky shores, wave-cut cliffs, and relatively short pocket beaches (e.g., coasts of western North and South America). Coastlines on the trailing edge of the continental plate are dominated by coastal plains that contain abundant estuaries and barrier islands (e.g., east coast of the United States). Modified after Davies (1973, Fig. 3) and Inman and Nordstrom (1971).

continental plate, which is moving westward at a rate of around 4 inches/year (Monroe and Wicander, 1998), broad deltaic and coastal plains predominate. To use this simple example, the western shoreline of the South American continent, and the North American continent for that matter, is dominated by **young mountain range coasts** characterized by: 1) high mountains caused by relatively recent orogenic (mountain building) activity; 2) bedrock of variable age but dominated by relatively young sedimentary and volcanic rocks; 3) active tectonic uplift (e.g., with old beach lines lifted up the sides of the mountains thousands of feet in some places); and 4) mostly short, high-gradient (steep) rivers emptying into the sea. The eastern sides of the two continents are dominated by depositional coasts characterized by: 1) coastal zone made up chiefly of broad coastal and deltaic plains; 2) bedrock composed generally of Tertiary and Quaternary sediments (young geologically speaking); 3) tectonically subsiding areas (e.g., thousands of feet of sediments have accumulated in some areas); and 4) many large and long rivers emptying into the sea. It is obvious, then, that the South Carolina coast belongs in the general category of depositional coasts. In fact, it is one of the prime examples, with its wide coastal plain, delta system, and network of rivers delivering sediments to the sea. Because the coast is located on the usually benign trailing edge of the continental plate, one does not usually associate dramatic tectonic events with such settings. However, earthquakes have occurred in the Charleston region from time to time.

Because the South Carolina coast is classified as a depositional coast, the rest of this discussion on the controls of shoreline types will focus only on that type of shoreline. Accordingly, the factor that exerts the most control on the geomorphological nature of a depositional coastline is the interaction of water in motion with the coastal sediments – typically called the **hydrodynamic regime** (*hydro* = water; *dynamics* = kinetic energy of the water). For purposes of interpreting the morphology of coastal features, the hydrodynamic regime is commonly expressed as the ratio of wave energy to tidal energy (i.e., how big the waves are versus how large the tidal range is). As noted in the Preface, during Hayes' early study of the world's shoreline for the U.S. Navy, he observed a striking correlation between tidal range (vertical distance between high and low tide) and shoreline characteristics of depositional coasts. Eventually, he classified depositional shoreline types based on this observation, differentiating between **microtidal** (tidal range less than 6 feet), **mesotidal** (tidal range = 6-12 feet), and **macrotidal** (tidal range greater than 12 feet) shoreline types. Surprisingly, he also noted a relatively equal distribution of the three classes of tidal range around the world's shoreline. Also, based on publications by the original *guru* of coastal geomorphology, W. Armstrong Price, he publicized two additional shoreline types, **wave-dominated coasts** and **tide-dominated coasts**. Generalized models of the basic types of depositional coasts are given in Figure 11.

As a generalization, most **microtidal depositional coasts** are wave-dominated and are characterized by:

1) Deltas with smooth outer margins and sand beaches (except in the case of huge delta systems on shorelines where the waves are of modest size, like the Mississippi Delta);
2) Abundant long barrier islands;
3) Sediment grain size distributions with coarser sediments onshore and finer sediments offshore, because of the role of waves (Breaking waves keep fine-grained sediments in suspension in the nearshore zone.); and
4) Multiple nearshore "breakpoint" bars offshore of the barrier island's beaches (Figure 11A).

In North America, wave-dominated coasts occur on much of the shoreline of Texas, the Outer Banks of North Carolina, and in Kotzebue Sound, Alaska.

On the other hand, most **macrotidal depo-**

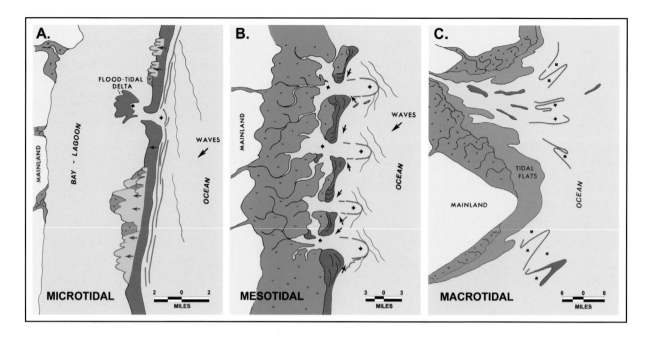

FIGURE 11. Generalized models of the three basic types of depositional coasts. (A) Microtidal (tidal range less than 6 feet, usually wave-dominated), (B) Mesotidal (tidal range = 6-12 feet, usually mixed energy), and (D) Macrotidal (tidal range greater than 12 feet, usually tide-dominated). The red lines in A represent nearshore, subtidal wave-formed sand bars, and those in B and C indicate intertidal sand deposits. The South Carolina shoreline is a classic example of a mesotidal (mixed-energy), depositional coast.

sitional coasts are tide-dominated and are characterized by:

1) Open mouthed, multi-lobate river deltas;
2) No barrier islands;
3) Sediment grain size distribution with coarser sediments offshore and finer sediments onshore, because of the role of tidal currents (In the more offshore areas, tidal currents are strong enough to keep fine-grained sediments in suspension, whereas near the high-tide line, weaker currents allow the fine-grained sediments to settle out. Because of the wide intertidal zones, waves break further offshore, thus playing a minor role in sediment distribution in the intertidal zone.);
4) Extensive tidal flats and coastal wetlands; and
5) Offshore linear tidal sand ridges (Figure 11C).

In North America, tide-dominated coasts occur in Bristol Bay and Cook Inlet, Alaska and in the northern reaches of the Gulf of California.

To complicate matters a little bit, most **mesotidal depositional coasts** have intermediate-sized tides and waves, thus another class of coast, **mixed-energy coasts**, has been recognized by us. Most mixed-energy depositional coasts are characterized by:

1) Short, drumstick-shaped barrier islands;
2) Numerous tidal flats and coastal wetlands; and
3) Very complex sediment distribution patterns (Figure 11B).

The South Carolina coast is the classic example of a mixed-energy coast. Other areas in North America where mixed-energy coasts occur are the Georgia coast and the Copper River Delta shoreline of Alaska.

A second major factor that determines certain key characteristics of depositional coasts is the regional **climate**. Wind patterns along a coast and offshore storms determine the average size of the waves, which has a major input in shaping the coast, as discussed earlier. Knowing whether or not the coastline is affected by hurricanes, extratropical cyclones (e.g., northeasters), or monsoons is key to understanding the seasonality of the variability of wave conditions along any given coastline. In South Carolina, hurricanes generate by far the largest waves that occur on the beaches in the state, but extratropical cyclones are also important at times. In certain microtidal areas, such as the south Texas coast, wind-driven tides control the geomorphological and ecological conditions of the coastal lagoons. The type of sediments on tidal flats and beaches is strongly related to the regional climate. For example, with regard to beach sediments, in tropical areas they are commonly composed of coral and algal fragments, carbonate precipitates and shell. In temperate regions, quartz sand is usually the dominate sediment type, with rock fragments and feldspar being abundant in sand near river mouths and along coasts with eroding bedrock. Other obvious climatic influences include:

1) The presence or absence of major rivers;
2) Effects of glaciers, freeze/thaw, and presence of permafrost in polar regions;
3) Types of weathering (chemical or mechanical) that produce the sediments that make up the coastal systems; and
4) Presence of beachrock in the tropics and permafrost in polar regions.

Although tides and waves determine the general framework of a depositional coastline, the climate determines the type of vegetation that occurs on the surface of the coastal habitats. Intertidal zones sheltered from waves are commonly populated by mangroves in the humid tropics, salt marshes in more temperate regions, and algal mats and evaporites in the arid tropics. As noted earlier, the South Carolina coast has a humid subtropical climate, thus the intertidal areas contain abundant salt marsh communities.

Another control on the nature of depositional coasts that is somewhat overrated in our opinion is the **source and supply of sediments**. Without such a supply, there would be no depositional coasts. On the South Carolina coast, the sediments at the shore are derived from rivers, erosion of the continental shelf, and erosion of areas elsewhere along the coast (most commonly to the northeast).

As shown in Figure 12, there are three major classes of sediments found on shorelines – gravel (particles > 2 millimeters (mm) in diameter, which consists of four separate classes, culminating with the largest - boulders), sand (particles between 0.0625 and 2 mm in diameter, ranging from very fine- to very coarse-grained), and mud (particles < 0.0625 mm in diameter and consisting primarily of

General Class	Wentworth Scale (Size Description)		Grain Diameter d (mm)
GRAVEL	Boulder		
			256.0
	Cobble		
			64.0
	Pebble		
			4.0
	Granule		
			2.0
SAND	Sand	Very Coarse	
			1.0
		Coarse	
			0.5
		Medium	
			0.25
		Fine	
			0.125
		Very Fine	
			0.0625
MUD	Silt		
			0.00391
	Clay		
			0.00024
	Colloid		

FIGURE 12. Definition of sediment types. From Wentworth (1922).

silt and clay). On the South Carolina coast, most of the sediment on the outer shore is fine-grained sand, whereas mud is by far the most common sediment type in the estuaries.

The last two factors affecting the nature of depositional shorelines are **local geological controls** and **sea-level rise**. Examples of geological controls include such factors as glacial effects, faulting, and variations in shoreline bedrock type. Along the South Carolina coast, local geological effects are more subtle, such as the relatively minor uplifting of the Cape Fear Arch to the northeast (Figure 4) and the gentle downwarping of the subsurface rock layers between Charleston and the South Carolina/Georgia border (the "Low Country"). With regard to sea-level rise, as noted earlier, sea level on the South Carolina coast has been relatively constant within the past 4,500 years, though some minor fluctuations have left their mark (Figure 9). A key question at the moment is will sea level continue to rise as it has been in the past few years, and, if so, how much? This issue will be discussed in more detail later.

4 COASTAL PROCESSES

TIDES

As shown in Figure 13, when the sun and moon are in line (full moon and new moon; **syzygy**), the gravitational attraction of the moon and sun are combined, giving rise to the maximum tides of the month (called **spring tides**). Minimum tides (called **neap tides**) occur during the first and third quarter of the moon (quadrative), when the two forces work in opposition. During some lunar cycles, the tidal range during spring tides can be almost twice that during neap tides. Because of its location so far from the earth, the gravitational pull (tide-raising force) of the sun on the world ocean is only 0.455 that of the moon. Therefore, the moon has the greater influence on the tidal cycle, controlling the timing of high and low tides.

In the much-simplified diagram shown in Figure 13, during syzygy two bulges of the high tide occur on opposite sides of the earth, one facing toward the moon and another one facing away from it. The bulge facing the moon is the result of the gravitational pull of the moon and sun. The bulge on the opposite side is much more difficult to explain. In actuality, the earth and the moon revolve together around the center of mass (*barycenter*) of the combined earth-moon system. The barycenter is displaced a distance from the center of the earth toward the moon, and it is always located on the side

of the earth turned momentarily toward the moon, along a line connecting the individual centers of mass of the earth and moon (Wood, 1982). To put it simply, as this system revolves around the barycenter, centrifugal force comes into play. This force is defined as the apparent force, equal and opposite to the centripetal force, drawing a rotating body away from the center of rotation, caused by the inertia of the body. Thus, the combined earth-moon system acts as a lever as the two equal and opposing forces revolve about a fulcrum, the barycenter. Therefore, the gravitational attraction of the moon (and sun) creates the bulge facing the moon and centrifugal force creates a bulge of relatively equal size on the opposite side of the earth.

The diagram in Figure 13 shows that, as the earth rotates about its axis, a given beach location on the earth's surface will pass by the peak of the two bulges every 24 hours and 53 minutes. The extra 53 minutes is required because that is how long it takes that beach location to "catch up" with the moon, which revolves all the way around the earth in about 28 days. Because of the intervening continents, a single bulge cannot pass all the way around the earth. Instead, individual **amphidromic systems** of different sizes are set up within the world's ocean basins. These systems are separate, more-or-less circular, gyrating tidal systems (standing waves) with a nodal point in the middle. This complicates

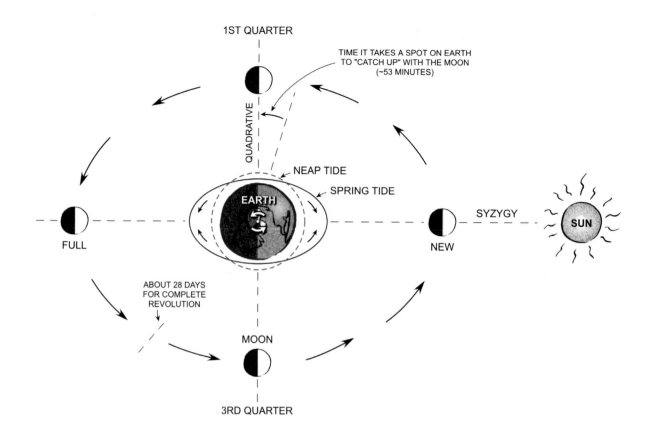

FIGURE 13. Forces involved in the generation of tides.

the story, but the primary forces just described above that cause the tides and are illustrated in Figure 13 still provide the basic explanation of the process.

There are many other complications to the simple story presented in Figure 13. For example, because the moon revolves in a slightly elliptical, rather than a circular, orbit around the earth, "the distance of the moon from the earth may vary between extreme limits separated by some 31,278 miles" (Wood, 1982). When the moon is closest to the earth, this "close encounter" is called the *perigee,* which is 13% closer than when it is furthest away (called the *apogee*). Thus, perigean tides may be 20% larger than normal mean tides. Furthermore, if the perigean tide corresponds with the syzygy (full and new moons), the so-called *perigean spring tides,* an increase of the tidal range of as much as 40% might be created (Wood, 1982).

The tides on the South Carolina coast are **semi-diurnal** (Figure 14), because two complete tidal cycles occur within 24 hours and 53 minutes, as opposed to some other areas in the world that have only one high and one low tide a day (**diurnal** tides; e.g., the Texas coast during some parts of the month). Therefore, low and/or high tide is about one hour later each day on the South Carolina coast.

The tidal range, as illustrated in Figure 14, is the vertical distance between high and low tide. The spring tidal range in South Carolina varies from about 5.5 feet at the North Carolina border to 8.2 feet at the Georgia border. This places much of the South Carolina coastline in the mesotidal class.

Whether the average tidal range in an area is large or small is dependent upon several factors, with the largest tides occurring where: 1) the water body has on open connection with the world

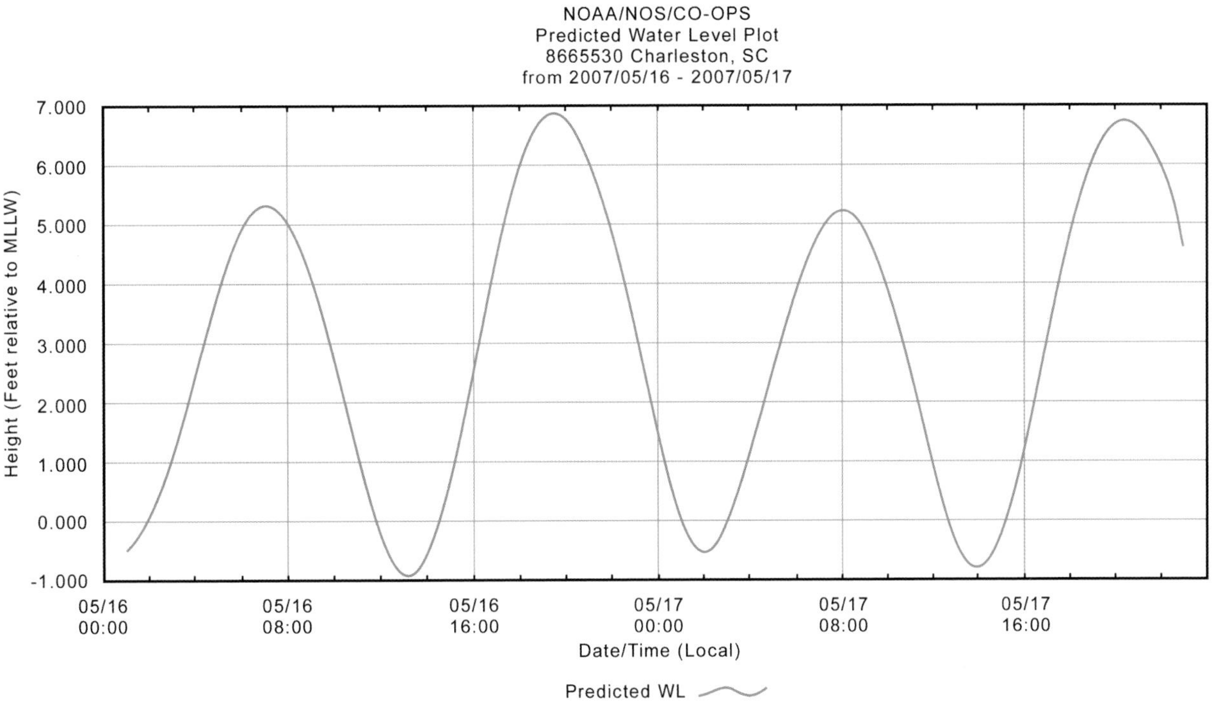

NOAA/NOS/CO-OPS
Predicted Water Level Plot
8665530 Charleston, SC
from 2007/05/16 - 2007/05/17

FIGURE 14. Typical tidal curve for South Carolina, showing two days. Note that there are two high tides and two low tides per day (plus about an hour), and each elevation is somewhat different. Thus, there is a "high" high tide and a "low" high tide each day.

ocean; 2) the continental shelf is wide; and, most importantly, 3) along shoreline embayments.

Other factors such as the hydrodynamic forcing of oceanic currents, position within an amphidromic system, and presence of resonant standing oscillation play a role in some areas, but none of these are as relevant as the first three factors with respect to the tides on the South Carolina coast. The largest tidal range along the southeastern coastline of the United States occurs at the head of the Georgia Bight (Figure 15), which is why the tides are so large at the South Carolina/ Georgia border. With the exception of the Bay of Fundy, an area where resonance plays a role, most, if not all, of the larger tidal ranges around the world occur in shoreline embayments, such as the Georgia Bight. Komar (1976) gave the following verbal explanation for this – "*as the tidal front approaches a narrowing indention of the coastline, …the enveloping shores constrict its movement and wedge the water together.*"

Currents generated by the tides are very strong at the entrances of the numerous tidal inlets that occur along the South Carolina coast. In some of the major inlets, current velocities may reach 3 or 4 knots [one knot = one nautical mile per hour (nautical mile = 6,076 feet)]. Tidal currents are also strong in the major tidal channels scattered throughout the estuaries and other backbarrier regions along the coast.

WAVES

Water waves are generated by the transfer of kinetic energy to the water surface as the wind blows over it. As shown in Figure 16, a typical wave is defined by its **wave length** and **wave height**. The length of time it takes two succeeding wave crests to pass a single point is called the **wave period**. In South Carolina, the average wave height is around 2 feet, but waves exceed this by several feet during a typical storm. During hurricanes, waves in the open

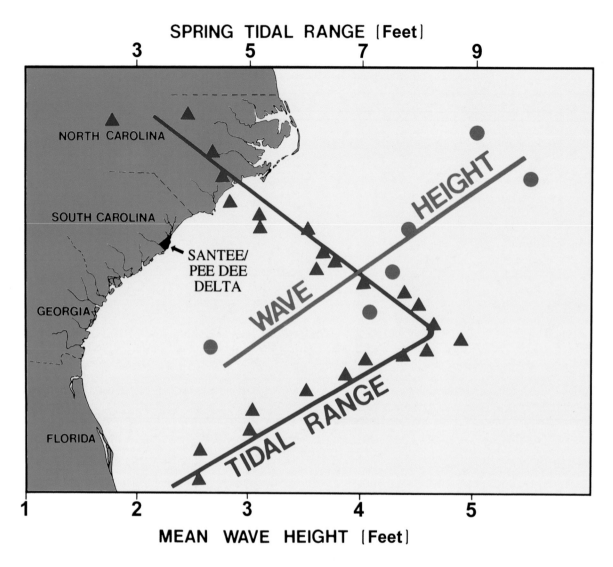

FIGURE 15. Variation of the tidal range and wave height along the shoreline of the Georgia Bight. Tide data from NOAA tide tables and wave data from Nummedal et al. (1977). Note that the smallest waves and largest tides occur at the head of the Georgia Bight (near the South Carolina/Georgia border).

ocean with heights greater than 50 feet have been observed. The factors that determine the height of waves include:

1) The fetch of the wind (length of water surface over which the wind blows to form the wave);
2) Wind velocity;
3) Duration of the wind; and
4) Incident swell.

As indicated in Figure 17, the waves in the area where the wind that forms them is blowing are referred to as **seas.** Once the waves pass out of the area of formation, they are referred to as **swell.** The differences between seas and swell are evident to even the casual observer, in that the water surface is choppy under sea conditions and smooth in areas where swell conditions predominate. Also, sea conditions are usually indicated by the presence of whitecaps, whereas whitecaps are not typically present on pure swell-type waves. Under sea conditions, the sea surface is irregular and the waves are asymmetrical (two sides not balanced), whereas the surface of swell is regular (relatively

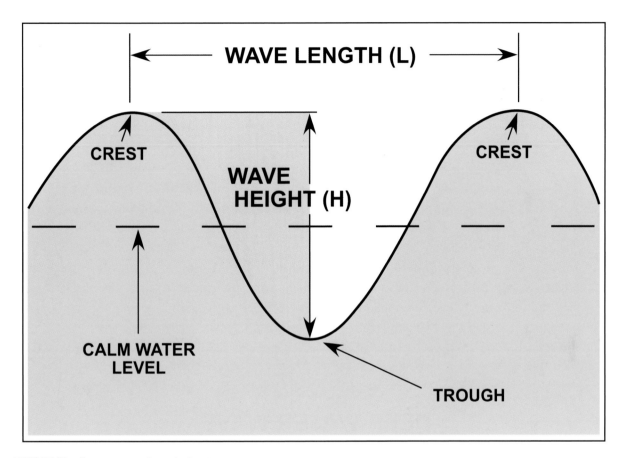

FIGURE 16. Components of a typical water wave.

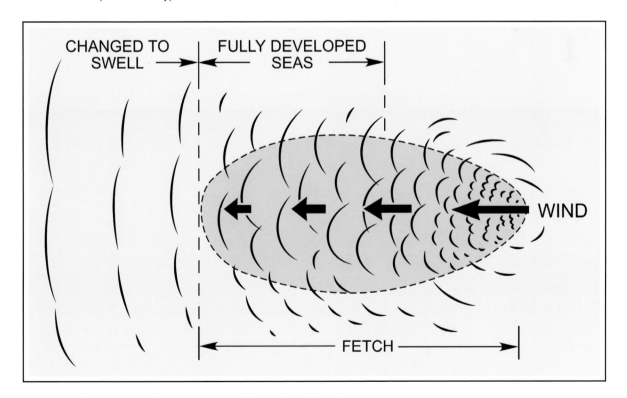

FIGURE 17. General pattern for the generation of waves by wind action.

smooth) and the waves are symmetrical. An observer standing on a beach may have difficulty distinguishing between the two types of waves, and it is very common for both types of waves to interact to form the resulting waves in the surf zone. To identify sea conditions, note the direction the wind is blowing on that day and look for the presence of whitecaps. Also, the period of swell waves is typically considerably longer than those waves generated by local onshore winds. Therefore, swell waves are usually formed by offshore storms that generate large waves. The waves generated by the storms move away from their zone of formation at a certain velocity with longer waves moving faster than shorter ones. A classic example of this escape of long waves from their zone of formation is the forerunner waves that precede the landfall of hurricanes. Commonly, surfers flock to the shore to take advantage of these huge, long-period "forerunner" waves.

As waves approach the beach from offshore, they increase in steepness and break in different ways to be discussed below. To understand why this happens, consider a cork on the water surface as a wave passes by. The cork ascribes a circular motion as the wave passes, in the process moving a small distance in the direction the wave is moving. As shown by the diagram in Figure 18, the water directly beneath the wave also undergoes a circular orbit, with the size of those orbits decreasing exponentially with depth until completely ceasing to orbit at a depth of about one half the wave length.

As the wave moves onshore, the wave begins to

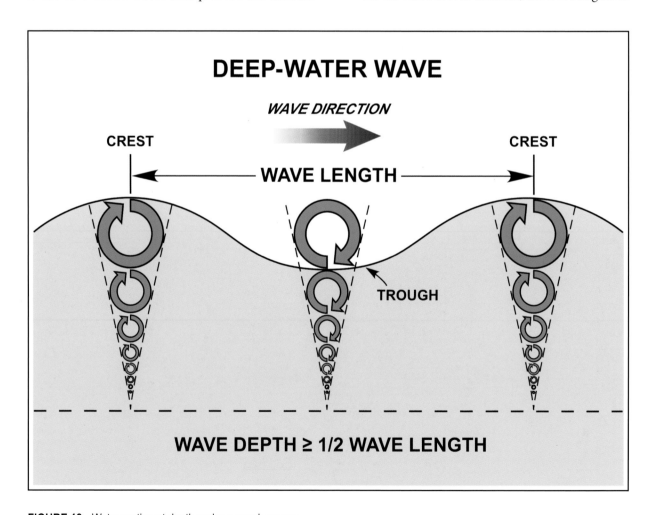

FIGURE 18. Water motion at depth under a passing wave.

"feel the bottom," that is, the orbital motion of the water under the wave begins to bump up against the bottom sediment where the water becomes **shallower than one half the wave length**. On the South Carolina coast, most storm waves, with the exception to those generated by major hurricanes, start to "feel the bottom" in water depths or around 20-30 feet. Thus, typical storm waves have lengths of up to about 50-60 feet. As the wave moves further landward, the orbits flatten out and become unstable, the waves get higher and the crests becoming sharper, with the velocity of the water in the top of the wave exceeding that at the bottom to such an extent that the surface of the wave eventually breaks. The zone where the waves are breaking is referred to as the **surf zone** (see Figure 19). After breaking, the water released by the broken wave impinges on the portion of the beach known as the beach face (also referred to as the **swash zone).** The water moving up the relatively steep beach face, known as the wave uprush, eventually loses its momentum and

that portion of the water that doesn't percolate into the beach face returns back down the slope under the influence of gravity (know as the backwash).

Breaking waves are usually of three distinct types – plunging, spilling, and surging. As pictured in Figure 20A, **plunging waves** take on a cylindrical shape (Hawaii Five-O type) and fall abruptly down with considerable force, usually breaking in the near vicinity of the beach face. The surfaces of **spilling waves**, on the other hand (shown in Figure 20B), start disintegrating into foaming lines fairly far offshore and continue to foam their way to shore, gradually decreasing in height as they go. **Surging waves** approach shore and wash directly up the beach face without breaking. As a generalization, for waves of any given height, the slope of the nearshore zone determines the type of breaking wave present. Plunging waves occur more commonly where the nearshore slopes are quite steep and spilling waves where the nearshore slopes are relatively flat. On the South Carolina

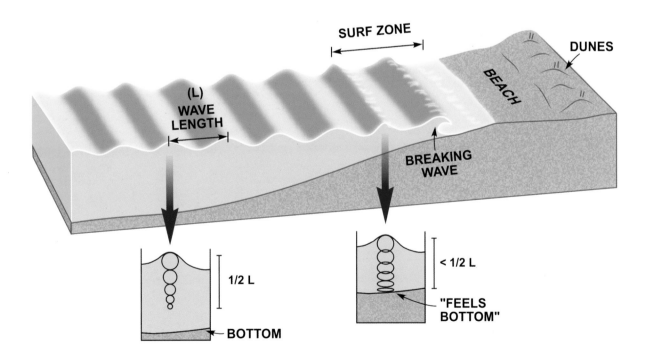

FIGURE 19. Characteristics of the surf zone under the influence of wave action. Note that in water shallower than approximately one half the wave length, the orbits of the wave-generated water motion under the waves (illustrated in Figure 18) flatten out as the wave begins to "feel the bottom." This causes the waves to slow down and become steeper and steeper until they eventually break.

FIGURE 20. Two types of breaking waves. (A) A plunging wave on the Mexican coast near Acapulco. The idiot running into the surf is co-author Hayes (summer 1961). (B) Spilling waves on the beach of Padre Island, Texas (summer 1979). In the upper right, two lines of waves are breaking on offshore bars.

coast, spilling waves are by far the most common type, with plunging waves mostly occurring during storms or under conditions of exceptionally large swell generated by offshore storms. This is no doubt due to the gentle nearshore slopes that characterize the South Carolina coast, because of its generally depositional nature. Under some conditions, a single wave may have the characteristics of both plunging and spilling waves. Surging waves are very rare along South Carolina's ocean coast.

When swell waves are viewed from the air, their crests are linear and parallel, forming what are known as wave fronts. In many situations, the wave fronts bend as they approach underwater features, this bending being referred to as refraction and diffraction. During the process of **refraction**, a segment of the wave front slows down as it enters shallower water (e.g., around an island or bar), with the outer edges of the front on either side of that segment moving faster, which allows the front to bend around the shallower feature (Figures 21A and 21B). In the purest sense, **diffraction** is defined as the lateral spread of wave energy along the wave crest from a point (an example is the wave formed when a pebble is dropped into a pool of still standing water). During incidents of wave diffraction near the shore, a segment of the wave front may speed up as it enters deeper water (e.g., into a channel) relative to the rest of the front, which allows that portion of the wave front to move more efficiently (faster) into the deeper water. As a result, an arc, or bow, develops in the wave front as it progresses into the deeper water.

When waves approach and break in the surf zone adjacent to the beach, they set up two very distinct, and in some cases very strong, nearshore currents – **longshore currents** and **rip currents**. As illustrated in Figure 22 (upper left and lower left), for a wave that approaches a beach at an angle, which most do, two types of water flow are generated that have the capability of moving sediment along shore. This type of **longshore sediment transport** is a key consideration with

regard to the stability of shorelines. An illustration of wave-generated longshore sediment transport is given in Figure 21A, which shows waves refracting around an island close to shore. The result of this refraction is the creation of a triangle-shaped sand body, called a tombolo, in the lee of the island at the point where two opposing sediment transport directions convergence. An example of this process on the California coast is shown in Figure 21B.

When waves come ashore at an angle, the wave uprush generated by the breaking wave flows at an angle to the slope of the beach face (Figure 22), which allows any sand grains moved by the shallow flow of the wave uprush to move obliquely up the beach face. On the other hand, the water in the backwash moves in a perpendicular sense directly down the slope of the beach under the influence of gravity. Therefore, with each wave sand is moved along the shoreline a certain distance, depending upon the size of the wave and its angle of approach (bigger waves and higher angles increase the distance of movement of the sand grains). Thus, the sand grain moves along the shore in a sawtooth pattern called **beach drift** (lower left diagram in Figure 22). This type of sediment transport is restricted to the beach face itself. At the same time, the obliquely approaching wave is piling up water in the surf zone that flows away from the direction of the approaching wave, thus generating a current that runs parallel to the shore (called the **longshore current**). On many shorelines, the longshore current is somewhat restricted to a trough that separates the beach face from a nearshore bar. These longshore currents can be quite strong, commonly exceeding a knot or two, and, in concert with the beach drift, play the major role in transporting sediment along the shore.

Another mode of longshore sediment transport occurs when evenly spaced bulges of sand along the beach face move down the shore like the sinusoidal loops one might throw along a rope. These bulges of sand, called **rhythmic topography** (see photograph in the upper right of Figure 22), are typically spaced

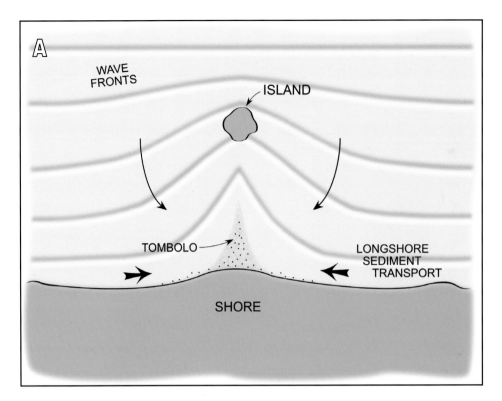

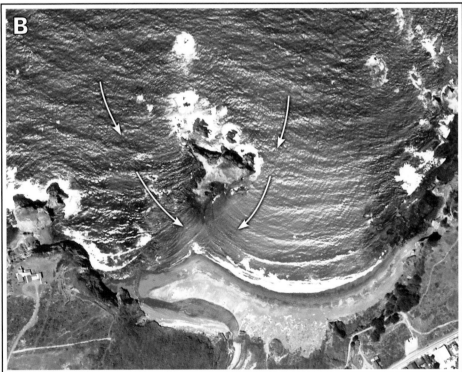

FIGURE 21. Illustrations of wave refraction. (A) Wave fronts bending around an offshore island, generating two opposing directions of sediment transport in the lee of the island. Where the two directions of transport meet, a triangular offshore projection of sand called a tombolo is formed. (B) Wave refraction around Gunderson Rock near Elk, California. The arrows, which indicate the direction of motion of the waves, are oriented perpendicular to the wave crests, or fronts. Courtesy of the California Spatial Information Library.

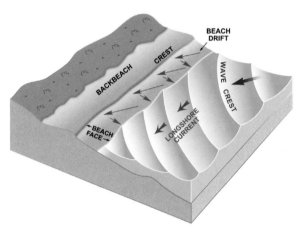

FIGURE 22. Longshore sediment transport. (lower left) General model for the generation of *beach drift* and *longshore currents* by waves approaching the beach at an angle. Under these conditions, the sand grains move in a sawtooth pattern along the beach face, and a strong longshore current is developed just off the beach face. (upper left) View looking north along St. Joe spit in northwest Florida showing waves approaching the beach obliquely out of a southerly direction. This high-tide picture shows the swash of the waves across the beach face. Arrows indicate the direction of the longshore currents generated by the obliquely approaching waves. Note waves breaking on the offshore bar with strong longshore currents generated in the trough on the landward side of the bar. Photograph was taken in April 1973. (upper right) Rhythmic topography along the shoreline of Matagorda Bay, Texas. Arrows indicate direction shoreline rhythms are moving through time, also under the influence of southeasterly waves. The water was relatively calm on the day this photograph was taken in August 1979. The white sediments are coarse-grained shell material (mostly oyster shells).

10 to 100s of feet apart. They are created by waves that approach the beach at an oblique angle. The side of the sand bulge facing directly into the wave approach direction orients perpendicular to that direction, with the beach face being aligned parallel with the approaching wave crests, the most stable configuration possible. However, the distance the sediment on the exposed beach face of the bulge can extend offshore is limited by water depth in most situations. Sand is therefore either lost from

the outer edge of the bulge into deeper water or is transported to its lee side. The result of this process is that the exposed face of the bulge erodes while sand accumulates on its sheltered side. Therefore, the bulge moves along the shore in what is known as the downdrift direction (direction in which the individual sand grains are moving along the shore within the longshore transport system), as indicated in Figure 22 (upper right). These features move at different rates depending on the size and angle of approach of the waves. Rhythmic topography does occur on the coast of South Carolina, but it is typically quite subtle. It is much more prevalent and conspicuous along shorelines with deeper offshore areas and bigger waves, such as at Cape Hatteras, North Carolina.

Maximum rates of longshore sediment transport on most any shore takes place during major storms. Two types of cyclonic storms, extratropical cyclones (northeasters) and tropical cyclones (hurricanes), impact the South Carolina coast. When the centers of these storms pass offshore of South Carolina on the way north, the dominant winds striking the coastline come out of the northeast. Therefore, although the details of the cause and effect are a somewhat more complicated story than this simple explanation, the dominant transport direction of beach sand on the state's shoreline is from northeast to southwest. A striking geomorphological indicator of such transport, a recurved spit on the southwest tip of Kiawah Island, is shown in Figure 23.

Research on the longshore sediment transport rates in South Carolina show that an average of around 150,000 cubic yards of sand moves from northeast to southwest along the sand beaches in the state in any given year (Kana, 1976; 1979). Because this number is a little difficult to grasp, we once calculated that this volume of sand would go a long ways toward filling Williams Bryce Stadium, where the University of South Carolina football games are played, and which, at the time of this writing, seats over 80,000 people. One hundred and fifty thousand cubic yards does seem like a large number, but in fact, it is rather puny by global standards. Places on the California coast have rates at least three times as much, and the Ghana coast in West Africa has rates that exceed one million cubic yards of longshore sediment transport in one year, as a result of the large waves on these exposed, open ocean coasts.

Rip currents (Figure 24) form most commonly along shorelines where waves come straight on shore

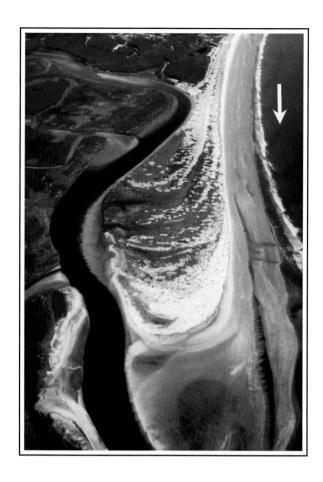

FIGURE 23. The recurved spit at the southern end of Kiawah Island, South Carolina at low tide on 10 June 1976 (infrared photograph by D.K. Hubbard). Because of the consistent movement of sand along the shore from northeast to southwest (see arrow; view looks northeast), the tidal inlet just beyond the bottom of the photo (Captain Sams Inlet) is forced to migrate to the south at rates of around 200 feet/year. As a result, the waves moving sand along the beach and into the inlet produce a curve in the beach that extends much of the way around the end of the island on the inside of the inlet. As this pulsating process continues, the spit continues to migrate to the southwest with the curving beach ridges marking the different stages of this advancement. Around 1948, a new inlet channel was formed at the narrow neck of the recurved spit, after which the inlet resumed its unceasing migration to the southwest. Therefore, the entire spit shown in this photograph, which was more than 5,000 feet long at that time, had formed in the previous 28 years.

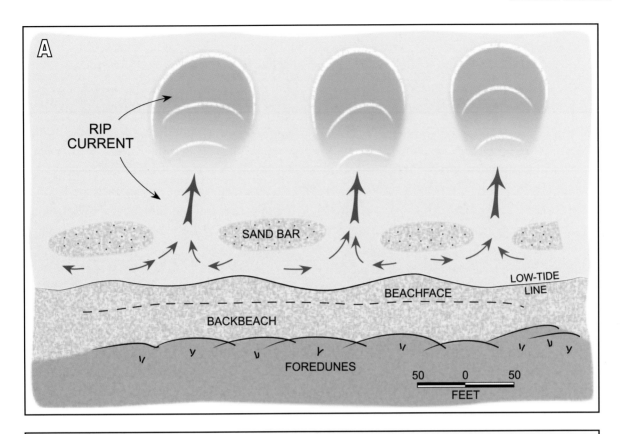

FIGURE 24. Characteristics of rip currents. (A) General sketch. (B) Rip currents along the shore of Cape Cod in April 1972. Note the onshore welding of sand bars in the space between the rip currents.

(not at an angle) with some degree of reflection off the beach face. These currents, which flow in a perpendicular direction straight off the beach, tend to be very regularly spaced. Based on extensive research, most rip currents owe their formation to the occurrence of edge waves – standing waves with their crests normal to the shoreline (Komar, 1976). Rip currents may also be associated with some types of nearshore bars and off certain man-made structures (e.g., groins, which are discussed below), but the most typical type is related to wave reflection and edge waves. It would take a Ph.D. in physical oceanography and at least a B.S. in physics to completely understand edge waves. At any rate, one of the authors of this book, Hayes, swears he has seen them with his own eyes off a steep beach in Ghana, but Michel is a non-believer in edge waves, because she says "she has never seen one." The only rip currents (not edge waves!!) the authors have actually seen on the South Carolina coast were generated by waves reflecting off the steep beach face at the southern end of Edisto Beach State Park. Tim Kana reports that these types of currents are "fairly common" in the Grand Strand compartment, but we have not visited that area as much as elsewhere. Rip currents no doubt also occur during the passage of major storms, and they can be very dangerous. If by chance you ever get caught in one while swimming off the beach, swim parallel to the beach so as to escape the current, which has the potential of carrying you far offshore.

HURRICANES

In terms of the scale of coastal dynamic processes, nothing can match those generated during the passage of a large hurricane. As it turns out, these types of storms have played a major role in the scientific careers of the authors of this book. For example, when hurricane *Carla* crossed the south Texas coast on 11 September 1961, Hayes, a graduate student at the University of Texas at the time, was well into his dissertation study of the

coastal environments of that area. In addition to doing extensive damage to the University of Texas field station at Port Aransas, a considerable amount of the data Hayes had collected up to that time was lost during the storm. Furthermore, the coastal environments themselves had been drastically changed by the hurricane. Ever the opportunist, Hayes turned his dissertation into a study of the geomorphological impacts of a major hurricane. This study became the first detailed investigation of such impacts on the coastline of the United States. Another hurricane, *Cindy* (1963), also provided input to this study (Hayes, 1965; 1967).

In 1979, the authors were investigating the pollution of the south Texas coast resulting from the *Ixtoc I* blowout in Mexico, the second largest oil spill in history. As the cleanup from that spill continued into the summer of 1980, hurricane *Allen* crossed the coast just north of Brownsville, Texas in early August. Happily, the storm cleaned the beaches of most of the remaining oil, thus we were able to bring that project to an end.

In September 1989, hurricane *Hugo* crossed the South Carolina coast just north of Charleston, doing a great deal of damage to the coast. We participated in some scientific studies of that hurricane, which became a kind of time line for references to the characteristics of the coast. The specific impacts of this storm are discussed in detail later.

In 2004, we moved to New Orleans to set up a companion business for our home office in Columbia. We had been there less than a year when hurricanes *Katrina* and *Rita* struck Louisiana. Our house in Uptown New Orleans suffered only very minor damage (it was above sea level), but the building that housed our 5th floor business office on the University of New Orleans campus was declared unlivable, because of an intense attack of mold. Therefore, we had to move the office to Baton Rouge, a 1.5-hour drive from our house. Meanwhile, Michel was working endlessly for NOAA and the U.S. Coast Guard on the over 120 oil spills in south Louisiana generated by the two hurricanes. After giving this living arrangement some consideration for a few months, we moved our business operation back to the main office in Columbia. At this point, we would be happy not to be involved with hurricanes any more for a while.

Technically, a **hurricane** is a storm of tropical origin with a cyclonic wind circulation (counterclockwise in the Northern Hemisphere) of 74 miles per hour (mph) or higher ($\geq$ 12 on Beaufort scale) (Dunn and Miller, 1960, p. 9). The "hurricane" is the North Atlantic member of the **tropical cyclone family** that includes the "typhoon" of the western North Pacific, the "cyclone" of the Indian Ocean, and the "willy-willy" of Australia (Tannehill, 1956). Tropical cyclones are the most powerful and destructive of all storms. Tornados have higher wind velocities (central axis velocities may attain 400-500 mph), but tropical cyclones are also intense (winds may reach 150-200 mph) and cover a much larger area. Extratropical cyclones (e.g., northeasters) are often larger than tropical cyclones, but do not achieve their intensity. Tropical cyclones exceed all other natural catastrophes in total damage to human life and property, and "no other weather phenomena receive such widespread and intense attention from the public" (Dunn and Miller, 1960, p. 7). Witness the public consciousness and concern, including intense media coverage, about hurricane *Katrina* (2005) and its remaining impact on New Orleans over two years after the storm.

Historically, hurricanes have occurred in the North Atlantic with a frequency of about 5 per year. In recent years, however, there has been a clear increase in the number and intensity of tropical storms and major hurricanes. Since 1995, the number of storms has risen dramatically, with an average at about 14 tropical storms, including about 8 hurricanes, per year between 1997-2006 (pewclimate.org/hurricanes). Like many scientific issues, the cause of this increase appears to be a matter of debate. Some prognosticators relate this increase to global warming, but others, most notably Professor William Gray of Colorado State University, a long-time predictor of hurricane frequency, believe "these multi-decadal variations are mostly due to changes in large-scale ocean circulations referred to as the Atlantic Ocean thermohaline circulation."

Not in doubt, however, is the fact that tropical cyclones occur most frequently during the months of August, September, and October. These months of maximum occurrence of hurricanes correspond roughly to the time when the Intertropical Convergence Zone, or equatorial trough, has its maximum divergence from the equator. Dunn and Miller (1960, p. 29) pointed out the importance of this divergence to hurricane formation in the following observation:

"When the Intertropical Convergence Zone is located within a few degrees of the equator only small vortices form along its course, but when it migrates northward, the influence of the rotating globe becomes great enough to transfer sufficient spin to converging currents to permit tropical cyclones to develop. In the Atlantic area, this occurs principally in the Cape Verde region and in the western Caribbean just north of the Isthmus of Panama."

The characteristics of hurricanes that produce the greatest geomorphological impact on shorelines are the storm surge and waves. Strong nearshore currents are also generated during the passage of a hurricane.

Not only is the **storm surge** the "primary cause of death and property damage during a hurricane" (Freeman et al., 1957, p. 12), but it is also the quality of hurricanes most responsible for making them such important geomorphological agents. The rise in water level brought about by hurricanes inundates vast areas of low-lying coastal regions, permitting widespread erosion and deposition of nearshore sediments, as well as catastrophic destruction of man-made structures.

The two most important factors in the generation of storm surges are the stress of wind on the sea surface (referred to as wind set-up) and the reduction of atmospheric pressure, or "inverted barometer effect." The theoretical relation of decrease in atmospheric pressure to water level rise is a 13-inch elevation of water level for each inch of mercury in pressure loss (Harris, 1963, p. 4). Winds of North Atlantic hurricanes have frequently reached velocities of 100 to 135 mph and a few have attained 200 mph (Dunn and Miller, 1960). The width of the path of a single storm may sometimes extend 200-300 miles. These factors, combined with the fact that a storm may last for several days, give hurricanes the ability to pile up tremendous quantities of water against the coastline.

Other factors that accentuate, or modify, storm surge are:

1) *Shoreline configuration* – Just as with astronomical tides, the funneling of surge waters into shoreline embayments, bays/lagoons, and estuaries will raise water levels considerably higher in their upper reaches compared to open coasts.

2) *Shape and slope of the continental shelf* – Greatest amplification occurs on a gently sloping shelf, like those off the Gulf Coast states and the Georgia Bight (Dunn and Miller, 1960).

3) *The angle the hurricane track makes with the coastline* – The counterclockwise circulation of the winds in a hurricane is such that, as the storm approaches the shore, only the winds on the right-hand side of the circular flow are moving in the same direction as the storm. Consequently, there is an additive effect of the wind set-up as the storm moves along only on the right side of the storm. Therefore, the peak surge usually occurs many miles "to the right" of where the storm's eye crosses the shore (Figure 25). If the storm strikes the shoreline at an acute angle between the right-hand side

of the storm track and the shore, the surge will be amplified even further. An example of this effect was the disastrous Galveston hurricane of 1900, in which more than 6,000 people died. That hurricane approached the northeast-southwest trending Texas coastline from the east-southeast, creating a maximum surge of 14.5 feet that washed over the low-lying barrier island (Conner et al., 1957). For more discussion of this storm see Erik Larson's book *Isaac's Storm: A Man, a Time, and the Deadliest Hurricane in History*.

4) *Stage of the astronomical tide (maximum effect at spring high tide)* – This is more important on shorelines with large tidal ranges.

5) *Rainfall effect* – Hurricanes may produce more rainfall in an area within a span of a few hours than is normally experienced in a whole year. This excess water may add considerably to flooding of coastal regions, especially if stream outlets are blocked by the storm surge.

6) *Several other less important factors.*

Probably the most spectacular geological effects produced by hurricanes are the result of erosion by breaking waves. The most impressive hurricane effect ever witnessed by the authors, who, as noted above, have studied the impacts of numerous hurricanes, was the absolute destruction of a coastal town called Holly Beach, Louisiana (all houses ripped completely off their foundations) by the destructive waves of hurricane *Rita* (2005).

A cubic yard of water weighs about three-fourths of a ton, and a breaking wave may move forward at speeds of 50-60 mph (Dunn and Miller, 1960). Their erosive effects are greatly increased when they ride the crest of a large storm surge, because of the extensive land areas and human developments that are exposed to erosion. Sometimes hurricane waves attain tremendous heights. It is common for wave heights of 40 to 50 feet to be observed offshore during major hurricanes.

Hurricane-generated waves that strike the

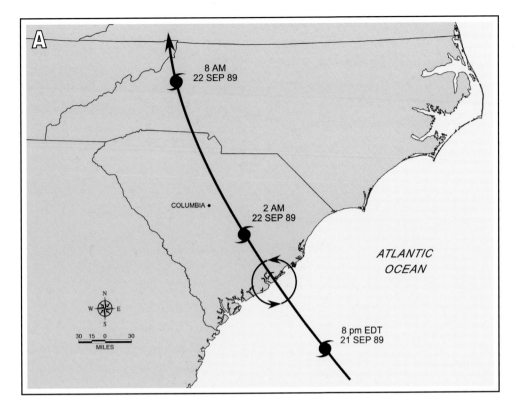

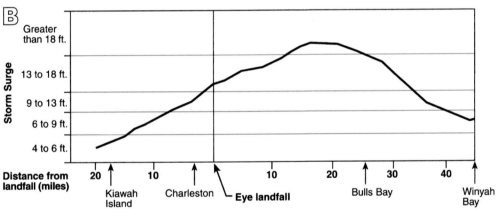

FIGURE 25. Path and storm surge of hurricane *Hugo* (1989). (A) Approach, landfall, and inland path of *Hugo* across South Carolina and North Carolina. (B) Plot of the storm surge associated with *Hugo*. Note that the peak of the storm surge (over 20 feet in places) occurred more than 20 miles north of where the eye crossed the shoreline. Twenty miles south of the eye's path, the storm surge was only about 2 feet. Modified after Lennon et al. (1996).

shoreline obliquely no doubt generate strong longshore currents, although they are almost never actually measured by scientists (wonder why?). Timbers and pilings from Bob Hall fishing pier on northern Padre Island, Texas, which was completely destroyed by hurricane *Carla* (1961), were found by Hayes for many miles south along the beach

after the storm. Exceptionally strong currents also flow through tidal inlets, as well as through new channels cut through barrier islands, during all major hurricanes.

Hurricanes cross the South Carolina and Georgia coasts much less frequently that they do the coasts of Florida, North Carolina, and Texas,

41

where, on the average, crossings happen more than once a year. This rarity of storm crossings in South Carolina appears to be the result of the facts that: (1) the shoreline at the head of the Georgia Bight is located so far inland of the general trend of the shoreline of the southeastern United States; and (2) hurricanes approaching that area tend in many cases to swing offshore in a more northeasterly direction, commonly clipping the barrier islands of North Carolina, which project far out into the ocean away from the general trend of the shoreline. The authors have been working on the South Carolina coast since 1972, and in those 35 years, the eyes of only a few tropical cyclones have crossed the coast – *David* in 1976, *Hugo* in 1989, *Gaston* in 2004, and a near miss by *Ophelia* in 2005.

EARTHQUAKES

Although not located in an area normally associated with such happenings, the earthquake that hit the Charleston area on 31 August 1886 was "the most destructive ever recorded in the eastern half of North America" (Murphy, 1995). The New Madrid, Missouri earthquake of 1811 was more intense, but not nearly as damaging to human-habitation structures because of the low density of population in that part of the country at that time. By the way, according to legend, the Madrid earthquake was caused by the great Shawnee warrior, *Tecumseh*, stomping his foot. Should you be interested in more details on that subject, we recommend the intriguing book, *A Sorrow in Our Hearts, The Life of Tecumseh*, by Allan W. Eckert.

Speaking of books, two books that give fascinating details on the Charleston earthquake are:

1) Chapter 9 in *Carolina Rocks!* by C. H. Murphy (1995), which gives some of the geological background; and
2) *City of Heros* by R. N. Cote (2006), which treats in great detail the human impacts of the

calamity.

The resident expert on the earthquake and other seismic activity in South Carolina is Dr. Pradeep Talwani of the Geology Department at the University of South Carolina. Talwani, (1982) explains why there are earthquakes in South Carolina.

The following, succinct summary of earthquakes in the Charleston area was taken from the web site earthquake.usgs.gov:

Charleston and its surroundings were devastated by a very large earthquake in 1886 (magnitude 7.3). Aftershocks, some of them large enough to be damaging by themselves, continued for years. Prehistoric earthquakes of similar size to the 1886 shock have occurred in coastal South Carolina at intervals of several centuries to several thousands of years. In recent decades, damaging earthquakes much smaller than that of 1886 have occurred every decade or two, most recently in 1995 (magnitude 3.5). Typically, smaller earthquakes are felt each year or two.
.....
Bedrock in the area is laced with faults that date mainly from the formation of the Appalachians and the birth of the Atlantic. However, in the Charleston area bedrock and its faults are buried beneath sand, silt, clay, and soft sedimentary rocks.

5 MAJOR LANDFORMS OF THE COAST

THE GEORGIA BIGHT

A nearly continuous chain of barrier islands lies along a broad shoreline arc bordering the southeastern U.S. states of North Carolina, South Carolina, Georgia, and Florida. This coastal area, referred to here as the Georgia Bight, is, in fact, the centerpiece for the longest single stretch of barrier islands in the world, the east and Gulf coasts of the North American continent – Cape Cod, Massachusetts to the Yucatan Peninsula, Mexico.

The shoreline of the Georgia Bight, which extends from Cape Hatteras, North Carolina to Cape Canaveral, Florida (Figures 4 and 15), is over 740 miles long. The Bight flanks the coastal plain on the trailing edge of the North American plate, which is generally tectonically stable to slightly downwarping (Inman and Nordstrom, 1971). Two northwest-southeast oriented tectonic highs (regions that historically are rising rather than sinking), the Cape Fear Arch on the north and the Ocala Uplift on the south, border the Bight (Figure 4). A corollary of the rise of the two arches is the downwarping of the underlying crust in the center of the Bight, in the area know as the "Low Country" in the southern reaches of South Carolina. This tectonic regime has apparently been in existence since late Paleocene (about 60 million years ago; Colquhoun et al., 1983). According to Colquhoun

et al. (1983), the bulk of the older sediments and sedimentary rocks that underlie the coastal plain of the Bight, which are Cretaceous to Holocene in age (a time span of approximately 100 million years), is only 900 feet thick at the North Carolina/South Carolina border, but thickens to 3,000 feet at the South Carolina/Georgia border (see Figure 6). This wedge of older sediments thins again further to the south.

Evidence published by Winker and Howard (1977) indicates that a zone of Pleistocene sediments deposited several 10s of miles landward of the present shoreline about 100,000 years ago, which they call the Chatham Sequence, has been raised 15 feet in North Carolina relative to the center of the Georgia Bight during that period of time. Thus, we can assume that the tectonic regime that originally formed the Georgia Bight (i.e., elevation of the two arches to the north and south and sinking of the crustal material underlying the center of the Bight), however mild, remains active even to this day. But don't worry too much about this, these numbers show that Charleston is probably sinking at a rate of about 0.2 inches per 100 years. This tectonic sinking should not be confused with relative sea level rise, which is a different issue altogether.

The Holocene coastline of the Georgia Bight is, for the most part, depositional, with barrier island, estuarine/bay/lagoonal, and minor deltaic and river deposits blanketing an irregular topography eroded during the low stands of the sea during the Pleistocene Epoch (Ice Age).

BARRIER ISLANDS

Introduction

Barrier islands are elongate, shore-parallel accumulations of unconsolidated sediment (primarily sand), some parts of which are supratidal (above the high-tide line most of the time, except during major storms). They are separated from the mainland by bays, lagoons, estuaries, or wetland complexes. According to Glaeser (1978), 76% of the barrier islands of the world occur along the coastlines of the trailing edges of continental plates and of large inland seas and lakes (e.g., Caspian and Black Seas). They do not occur on coasts with tidal ranges greater than about 12 feet, because their primary mechanism of formation, wave action, is not focused long enough at a single level during the tidal cycle to form the island, and the strong tidal currents associated with such large tides transport the available sand to the offshore regions. Some barrier islands do occur, primarily as spit forms, on leading edge and glaciated coasts, but they are a minority coastline type in those areas.

Types of Barrier Islands

Depositional, coastal-plain shorelines, such as the South Carolina coast, typically have barrier islands located between major river deltas and estuaries. Two major types of barrier islands are present on this coast, those that consistently migrate landward (**transgressive**) and those that build seaward (**prograding**). Although all barrier islands tend to change to some extent over time, the rates of landward migration and/or seaward growth are different from place to place. In some parts of the South Carolina coast, the landward-migrating barrier islands may move landward 10s

of feet per year. However, those that build seaward usually grow more slowly, at rates of less than 10 feet per decade. Which of these two types of islands are present at a particular location along the coast depends upon the rate of sea-level change relative to sand supply (Curray, 1964) at that location. Diminished, or low, sand supply and/or rapid sea-level rise promotes the development of landward-migrating islands, and *vice versa* for those that build seaward. There are 75 miles of prograding (seaward-building) barrier islands and 22 miles of landward-migrating (transgressive) barrier islands on the South Carolina coast, in a stratigraphic/geomorphological sense. That is, there are a number of locations along originally prograding barrier islands where the beach is eroding at the present time, because of diminished sand supply or other reasons, but the part of the island that remains was formed during the progradational phase of island development. Also, there may be short-term episodes of beach buildup on landward-migrating barrier islands where an unusually large amount of sand comes on shore.

As illustrated in Figures 26 and 27, landward-migrating barrier islands are composed of coalescing washover fans, or a washover terrace, that is overtopped at high tides, usually several times a year. In the process of migration, the entire washover terrace complex moves landward, leaving an eroded nearshore zone in its wake. As a result of this type of migration, in three dimensions the entire complex consists of a relatively thin (<6-12 feet) wedge of sand and shell of the washover terrace that overlies muddy sediment originally deposited in the lagoons or wetlands behind the islands (see cross-section in upper half of Figure 26). Generally speaking, if you are walking on the beach of a barrier island at low tide and you find marsh sediments exposed in the intertidal zone, that particular part of the barrier island is migrating landward. A good example of this that is readily accessible occurs along the northern end of the beach at Edisto Beach State Park. The islands

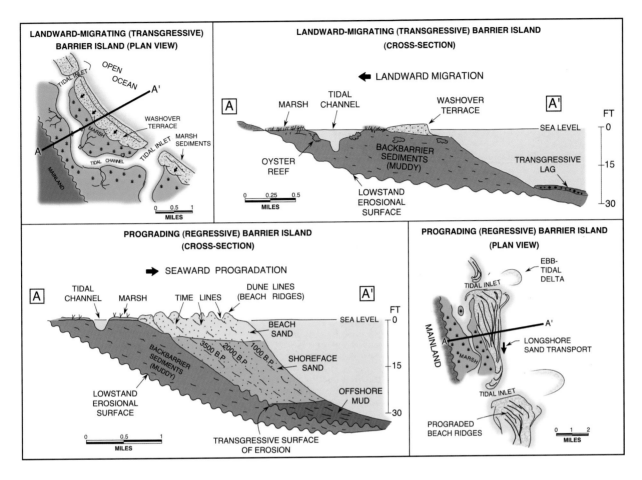

FIGURE 26. Morphology and subsurface three-dimensional configuration (stratigraphy) of prograding (regressive) barrier islands, in the bottom row; and landward-migrating (transgressive) barrier islands in the top row. The cross-sections (A-A') are located in each of the plan view maps.

on the outer margin of Cape Romain are also of this type, but they are accessible only by boat. Because of their continual landward migration, these types of islands are, needless to say, impractical sites for human development. The landward-migrating barrier islands in South Carolina are relatively short, 1-5 miles on the average, because new inlets are created where the migrating islands intersect tidal channels (see photograph in Figure 27).

Prograding (seaward-building) barrier islands (Figures 26, 28, and 29) are typically composed of multiple, relatively parallel linear ridges of sand topped by vegetated sand dunes that originally formed as front-line dunes on the backbeach (called **foredunes**). The most notable changes on

these types of islands occur where adjacent tidal inlets migrate into them or when the inlets expand dramatically during hurricane storm surges. As a result of this type of seaward growth, in three dimensions they consist of a wedge of sand 24-30 feet thick that has built seaward over offshore mud (see cross-section in the lower half of Figure 26). Most of the major developed barrier islands in the state, which typically are greater than 10 miles long, are of this type (e.g., Isle of Palms, Kiawah Island, and Hilton Head Island). When human development occurs on these types of islands, buildings are usually secure from all but the most extreme hurricanes, if they have been set back an adequate distance from the front-line dunes and

OLDER SHORELINES

FIGURE 27. Landward-migrating (transgressive) barrier islands at Edingsville Beach, South Carolina. This shoreline has retreated almost a mile since the 1850s because of the trapping of sand on the major ebb-tidal delta located north of the area. Note the new tidal inlet that was created by the intersection of the retreating washover terrace with a tidal channel (arrow). Photograph by M.F. Stephen taken in October 1974.

tidal inlets. But, that security will vanish if a major rise of sea level occurs (discussed further later).

The morphology of the prograding barrier islands longer than about 6 miles takes on a characteristic drumstick appearance, as shown in Figure 29. Two factors that enhance the development of the drumstick shape on the South Carolina coast are:

1) The occurrence of significant masses of sand in the form of large intertidal sand bars (swash bars) that pass around the outer margin of a large sand shoal (**ebb-tidal delta**) associated

with the tidal inlet on the northeast end of the island. These huge swash bars eventually move toward shore and attach to the beach (Figure 29A), building out the northeast end of the island accentuating the drumstick shape (see photograph in Figure 28); and

2) Refraction of the dominant northeasterly waves around the ebb-tidal delta, a process that creates a sediment transport reversal to the northeast in the area in the lee of the ebb-tidal delta. This relatively minor reversal from the normal northeast-to-southwest sediment transport direction that occurs along the rest of the island (see model in Figure 29A) allows some sand to remain in the inlet area, which accumulates on the northeast end of the island.

These two processes also combine to create major **downdrift offsets** at many of the tidal inlets in the state. **Drift** is a term engineers use for the volume of sand that typically moves along a beach during some specific interval of time. The term updrift refers to the direction from which the sand comes and downdrift is the direction in which it travels. Normally, one would expect the two barrier islands on either side of a tidal inlet to be oriented along the same general trend of the shoreline. When this is not the case, the adjacent islands are said to be offset at the inlet. Because of the two processes described for the formation of the drumstick shape of many of the prograding barrier islands in South Carolina, the islands to the southwest of most of the inlets are offset in a seaward direction. Illustrations of such downdrift-offset inlets in South Carolina are given in Figures 30 and 31. There are a few exceptions to this trend, which will be discussed later in the sections that deal with the individual compartments of the coast.

Origin of Barrier Islands

The information we have, based on many years of study of the Georgia Bight coastline, indicates

FIGURE 28. An oblique infrared view of Kiawah Island from the northeast, which clearly illustrates the drumstick configuration of this 8.5-mile-long, prograding barrier island. Note the presence of linear ridges of sand vegetated by maritime-forest vegetation (arrow), which indicate the positions of the foredunes just back of the beach at earlier stages in the growth of the island. Photograph by D.K. Hubbard taken on 18 March 1976.

that the barrier islands in this area originated in one of three ways:

1) Elongation of sand spits away from a headland with permanent segmentation of the spits as they grow as a result of the formation of permanent tidal inlets through them during storms, creating individual islands (Gilbert, 1989; Fisher, 1967; see Figure 32).

2) Formed at lower sea level and migrated to their present position (Swift, 1975); and

3) A process we call the transgressive-regressive interfluve hypothesis, which has been well documented by others (Figure 33; Pierce and Colquhoun, 1970; Moslow, 1980).

One of the major proponents of the **spit-elongation hypothesis**, John Fisher, cited the northern part of the Outer Banks of North Carolina as one of his examples of barrier islands that have been formed by that mechanism. The fact that some of the shoreline sand bodies that build away from the uplifting Cape Fear and Ocala Arches have a spit-like form is evidence that a spit elongation mode of origin could have accounted for some of the barrier islands in these tectonically uplifting areas as well. This mode of origin has been suggested for other areas in the world, for example, the islands on the central Texas coast (J. McGowen, pers. comm.).

In a paper written in 1975, the noted marine geologist Don Swift stated that barrier islands originated at a lower stand of sea level and migrated over the drowning coastal plain as sea level rose.

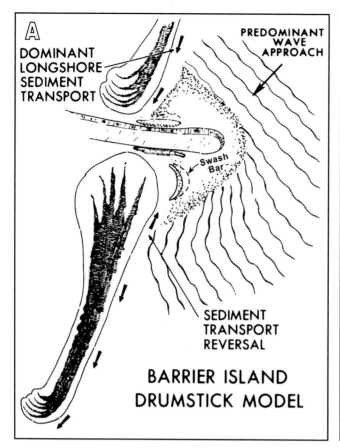

FIGURE 29. Typical morphology of prograding barrier islands in South Carolina. (A) Barrier island drumstick model, primarily the result of welding of masses of sand derived from the updrift ebb-tidal delta in the form of large swash bars. (B) Fripp Island, South Carolina, a 4.5 mile-long prograding barrier island that reflects the drumstick configuration. The updrift end of the barrier island is at top of the photograph. Drift is the term engineers use to define the volume of sand that is moved along the shore by longshore currents. **Updrift** is the direction the sediments are coming from and **downdrift** is the direction in which they are headed. Note intertidal bars along the beach. The multiple sets of forested beach ridges in the lower left represent different episodes of progradational barrier island growth during an earlier highstand of sea level (possibly as much as 2,000 years ago; see sea level curve in Figure 9). A minor drop in sea level preceded the highstand during which the developed part of the island was deposited (right side of the island in the picture, the part of the island that emphasizes the drumstick shape). Infrared photograph taken at low tide in December 1979.

This mechanism is probably the case in several localities in the Bight where the islands continue their landward migration even to the present day (e.g., Core Banks, North Carolina and just north of Cape Canaveral, Florida). He doesn't say exactly how the islands originated at that lower level.

The mode of formation for most of the larger prograding barrier islands on the South Carolina, such as Kiawah Island, is clear and well documented. Four major steps take place in this mode of formation according to the **trans-gressive-regressive interfluve hypothesis**, which is illustrated in Figure 33.

1) A narrow, landward migrating barrier island moved rapidly across what is now the inner continental shelf, leaving behind a thin lag of coarse material on top of an erosion surface across the continental shelf, called the transgressive surface of erosion;

2) The topography over which the shoreline advanced was irregular, and estuarine waters

48

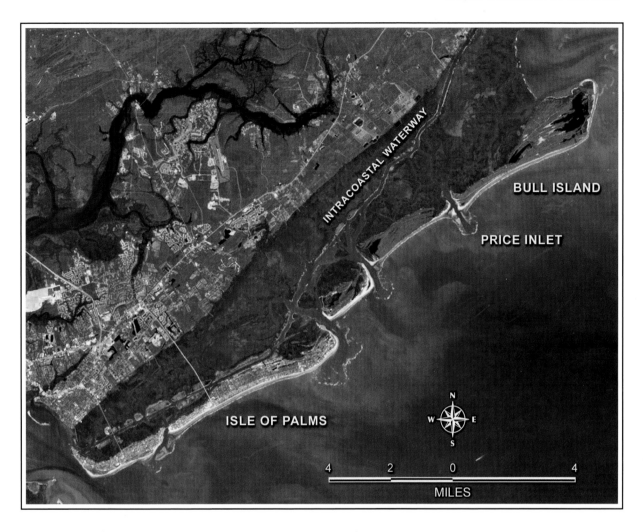

FIGURE 30. The section of the coast between Charleston Harbor (to the southwest) and Bulls Bay (to the northeast). Vertical infrared image acquired on 23 October 1999. Note the drumstick shape of the two larger islands, Bull Island and Isle of Palms. All of the inlets, except Price Inlet at the southwest end of Bull Island, have distinct downdrift offsets (Figure 31).

flooded the numerous river valleys formed when the shoreline was further offshore. Isolated, primary landward-migrating barrier islands, consisting of washover terraces composed of coarse-grained sand and shell, continued to develop on the exposed interfluves between the drowned valleys (top diagram in Figure 33);

3) When sea level stopped rising and a relative stillstand occurred about 4,500 years ago, shoals developed at the entrances of the estuaries created by the drowning of the valleys and a longshore sediment transport system was initiated along the face of the stranded

barrier islands. Over time, beach ridges began to develop, eventually impinging upon the adjacent estuary entrances (middle diagram in Figure 33). As a well-defined inlet throat evolved, a shoal off the entrance (ebb-tidal delta) formed around which sediment was bypassed, augmenting beach-ridge growth downdrift of the inlet (bottom diagram in Figure 33); and

4) As the barrier island matured, and minor fluctuations of sea level occurred, parts of some of the originally prograding beach ridges were eroded as a result of tidal-creek and tidal-inlet migration.

FIGURE 31. View looking southwest from the southwest end of Bull Island. Inlet in foreground is Price Inlet and the Isle of Palms is visible in the distance. The two tidal inlets in the middle distance, Dewees and Capers, show distinct downdrift offsets. Infrared photograph taken circa 1978.

The end result of all this was a prograding, drumstick-shaped barrier island, such as the ones illustrated in Figures 28, 29, and 30.

Over the years, we have observed the formation of features we called *micro-barriers* on tidal flats in Iceland, Alaska, and Cape Cod. Although exceedingly small (measured in a few inches to a couple of feet), these features contain many of the elements found on typical barrier islands, such as recurved spits, tidal inlets, lagoonal areas on their landward sides and so on. In all cases, these features formed within minutes to a few hours. The requirements for their formation was a nearly flat,

gently sloping surface suddenly flooded by shallow water with a strong enough wind to form breaking waves. If you accept the obvious problem of the differences in scale, this illustrates that features like barrier islands can form on relatively flat, flooded surfaces with just the right wave conditions. At the scale of normal barrier islands, the waves in question, formed during stormy periods, would have to break in the proper depth to create a bar on either the landward or seaward side of the breaking wave. A drop in the water level after the storm passes would allow the bar to emerge with a lagoon developing on its landward side. No doubt these conditions exist on depositional coasts in many parts of the world. The only remaining question is exactly what depth, slope, and wind/wave conditions does it take to form the island? Once that is figured out, it should be no surprise that long chains of barrier islands occur offshore of lagoons and estuaries, regardless of some of the other theories proposed for the origin of barrier islands.

TIDAL INLETS

As defined here, tidal inlets are **major channels that intersect barrier islands,** or separate two barrier islands, usually to depths of 10s of feet. An oblique aerial photograph of the tidal inlet that separates Kiawah and Seabrook barrier islands is shown in Figure 34. Tidal inlets are the focus of the most dynamic changes that occur on any component of a barrier island system. Because they connect the open ocean with more sheltered back-barrier settings, including some potential harbor sites, engineers have long been engaged in efforts to control tidal inlets, commonly by constricting their migration with a set of two parallel jetties. In South Carolina, three inlets have been so "improved" – at Murrells Inlet, and at the entrances to Charleston Harbor and Winyah Bay. Also, jetties have been built at the mouths of the Savannah River on the Georgia border and Little River on the North Carolina border.

SPIT-ELONGATION HYPOTHESIS

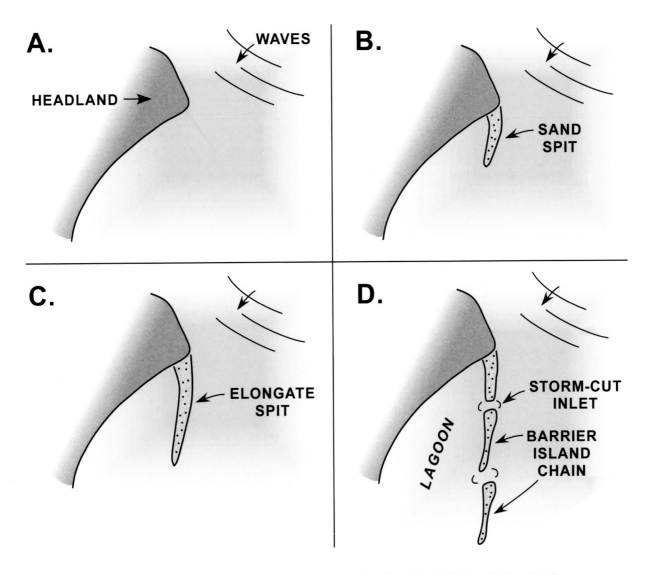

FIGURE 32. Spit-elongation hypothesis for the origin of barrier islands based on Gilbert (1889) and Fisher (1967).

In South Carolina, tidal inlets have been formed by one of three processes:

1) Storm-generated scour channels;
2) Closure of estuarine entrances by growth of sand spits; and
3) Intersection of large tidal channels in salt marshes by the landward migration of barrier islands.

When the storm surge of a tropical or extratropical cyclone floods a barrier island, water flowing through breaks in the dunes is capable of scouring channels across the island. This is a mechanism by which many new tidal inlets are formed (Hayes, 1965; Shepard and Wanless, 1971); however, most of these channels are closed off by wave-transported and wind-blown sand within a few weeks after the storm. Studies of the Outer

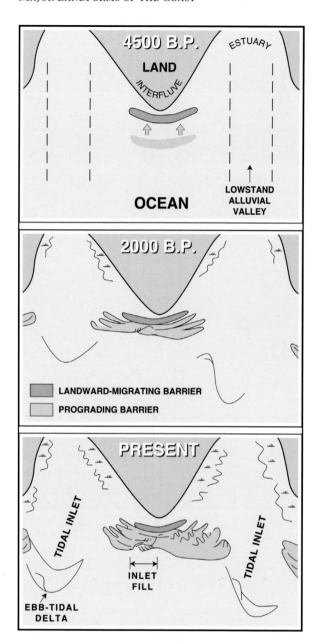

FIGURE 33. Model for the origin of the prograding barrier islands on the South Carolina coast – *transgressive-regressive interfluve hypothesis*. Based on Pierce and Colquhoun (1970) and Moslow (1980).

Banks of North Carolina by Moslow and Heron (1978) showed an abundance of tidal inlet-fill deposits preserved underneath a long, landward-migrating and wave-dominated barrier island (Core Banks). These deposits are the record of ephemeral inlets generated by scour during major hurricanes.

Captain Sams Inlet, pictured in Figure 34, was presumably formed by this mechanism during a hurricane in the mid to late 1940s. More details on the historical evolution of this inlet are given in the discussion of the Barrier Islands compartment.

The studies of the evolution of the barrier islands in South Carolina discussed above demonstrated that inlets are formed by the narrowing of estuarine entrances by the growth of sand spits once sea level reached near its present level around 4,500 years ago (as illustrated in Figure 33). Over time, the much narrowed inlet throats were stabilized near the center of the original valleys, achieving depths of as much as 90 feet in places. The throats of the large inlets in South Carolina formed by this mechanism are essentially anchored in bedrock, and our studies of historical trends of inlet changes (Zarillo et al., 1985) show that they do not migrate significantly.

The photograph in Figure 27 illustrates a new tidal inlet formed by the landward migration of a barrier island across a major tidal channel in a backbarrier marsh. One such inlet, which now is used by ocean-going shrimp boats, was formed on the outer shore of Cape Romain during hurricane *Hugo* (1989).

The satellite image of the central portion of the South Carolina coast given in Figure 35 shows the location of inlets formed by each of the three primary mechanisms: 1) Captain Sams Inlet cut by a major hurricane; 2) Stono and North Edisto Inlets formed by the partial closing off of the mouths of two large estuaries; and 3) a number of small inlets formed along Edingsville Beach (south of North Edisto Inlet) as a result of the landward migration of a barrier-island system.

The ultimate morphology, or shape, of a tidal inlet in a barrier island complex, like those on the South Carolina coast, is the product of the constant interaction, or contest if you will, between tidal and wave-generated forces. Longshore currents generated by wave action move sediment along shore in such volumes that would normally fill

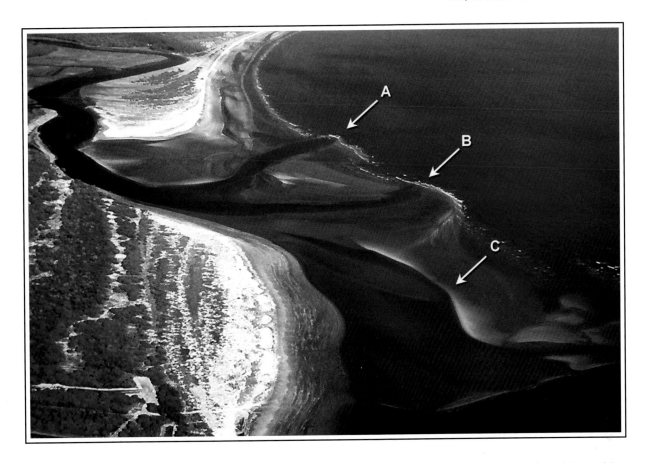

FIGURE 34. Low-altitude, infrared view of Captain Sams Inlet, South Carolina, a tidal inlet that migrates to the southwest (view looks northeast) at the rate of around 200 feet per year. The complex shoals seaward of the inlet are part of the ebb-tidal delta, which, at this time, contained three outlets. The outlet in the distance (arrow A) was cut by hurricane *David* in September 1979. The channel in the center (arrow B) eventually became the dominant channel in the spring of 1980. The bar in the foreground (arrow C) gradually moved landward, allowing a large packet of sediment to bypass the inlet and weld to the beach, a process called **bar-bypassing** (Sexton and Hayes, 1982). As a result of these types of bars welding to the beach, the downdrift offset of the beach south of the inlet developed. The light colored bulge of sand behind the beach (in the foreground) had been formed in this manner within the past few years before this photograph was taken (at low tide in December 1979). The bright red vegetation (wax myrtle – *Myrica cerifera*) was growing on older deposits.

the inlet in the absence of tidal currents. Sediment carried into the inlet throat (see general model in Figure 36) by the wave-generated currents is constantly swept away by tidal currents that scour the deep throat between the barrier island beaches (which encroach upon it from both sides). Sediment carried away by flood currents deposit a lobe of sediment on the landward side of the inlet called the **flood-tidal delta** and that carried seaward by the ebb currents form a seaward lobe called the **ebb-tidal delta**.

The general model of tidal inlets given in Figure 36, which shows relatively equal volumes of sediment in the two tidal deltas, is rare in nature. More typically, in microtidal, wave-dominated areas like the Texas Coast, the flood-tidal delta is much larger than the ebb-tidal delta because of the influence of wave action in both moving sediment inside the inlet and in erosion of the outer lobe (among other factors). On mesotidal, more tidally influenced coasts like South Carolina's, ebb-tidal deltas tend to be much larger than flood-tidal deltas.

FIGURE 35. Vertical infrared image of Kiawah Island and vicinity acquired on 23 October 1999, courtesy of Earth Science Data Interface at the Global Land Cover Facility. The locations of inlets originating by three different mechanisms are shown on this image. Captain Sams Inlet was cut most likely during a hurricane storm surge in the middle to late 1940s. Stono and North Edisto Inlets were formed by barrier islands converging over lowstand valleys, as is illustrated in Figure 33. A number of small tidal inlets were formed south of North Edisto Inlet as a result of landward-migrating barrier islands moving across major tidal channels (as illustrated in Figure 27).

In fact, flood-tidal deltas are almost non-existent on both the South Carolina and Georgia coasts. This strange occurrence has been studied in some detail. In the simplest terms, because of the complex nature of the backbarrier regions of the barrier islands in South Carolina (extensive marshes and tidal flats cut by numerous tidal channels), water "running uphill" into these areas during the flood portion of the tidal cycle takes longer to fill the backbarrier region with water at high tide than it takes that region to empty when the tide falls. This means that the flood phase is considerably longer than the ebb phase. With essentially the same amount of water

going in as comes out, much stronger currents are generated during the falling tide. These strong ebb currents are able to sweep sand brought into the inlet by the wave-generated currents out of the inlet and deposit it on the ebb-tidal delta.

The large ebb-tidal deltas on the South Carolina coast contain huge volumes of unconsolidated sand (77% of the total volume on the coastline; Hayes, Sexton, and Sipple, 1994; Sexton and Hayes, 1996). A general model of these prominent sand bodies is given in Figure 37. Compare the model with the examples of real tidal inlets given in Figures 38A and 38B. The morphological components of ebb-tidal

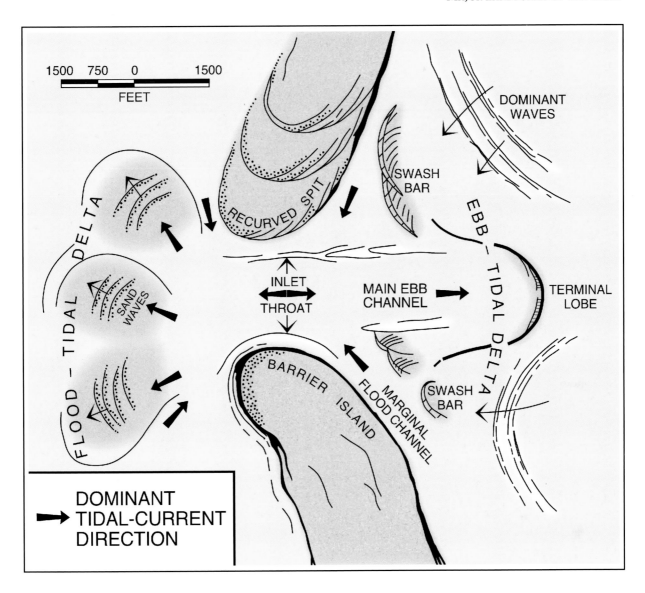

FIGURE 36. General model of the morphology of tidal inlets.

deltas include a **main ebb channel** (with slightly stronger ebb currents than flood currents) flanked by linear sand bars on both sides and a **terminal lobe** at the seaward end. The main ebb channel is bordered by a platform of sand (swash platform) dominated by swash bars (intertidal bars built up by wave action). The swash platform is separated from adjacent barrier beaches by **marginal flood channels** (which are dominated by flood currents). Figure 38A shows an example of one of these huge ebb-tidal deltas at the entrance to the North Edisto Inlet. This monster tidal delta, which extends 4.3

miles out into the ocean, contains 230 million cubic yards of sand. Note how well the shape of this ebb-tidal delta conforms to the general model given in Figure 37, which was originally proposed for the New England coast by Hayes and his students at the University of Massachusetts. That's the magic of physics! Water always runs downhill! Always conforming to Newton's Law of Gravity.

Property losses can occur on developed barrier islands, particularly those in mixed-energy settings, as a result of inlet migration or expansion. The effect of **inlet migration** on downdrift properties is well

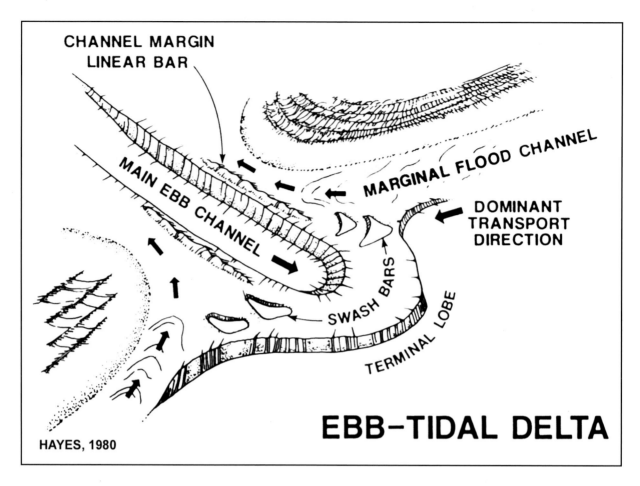

FIGURE 37. Typical morphology of ebb-tidal deltas in mesotidal settings (after Hayes, 1980). Arrows indicate dominant direction of tidal currents. This model was derived for the ebb-tidal deltas of the New England area, but it applies equally well to the inlets of South Carolina and other barrier-island systems around the world.

understood, resulting in remedial measures, such as inlet stabilization with jetties, or inlet relocation, as was done at Captain Sams Inlet, South Carolina in 1983 (explained in detail in the discussion of the Barrier Islands compartment). Studies on the central South Carolina coast following hurricane *Hugo* (1989) showed that several inlets expanded to accommodate the 12-20 feet storm surge associated with the hurricane. Price and Capers Inlets became 25 percent wider than they were before the storm; however, within a few months, they were back to their original width. The expansion of the inlets and scour of adjacent subtidal areas caused major property losses at the Isle of Palms, Pawleys Island, and Sullivans Island. The impacts of hurricane *Hugo* are discussed in more detail later. A generalized

model of inlet expansion, based on detailed study of historical charts, maps and aerial photographs, allowed us to establish a meaningful construction setback line for Kiawah Island, South Carolina before the major phases of development began in 1975 (Hayes, 1977). Kiawah Island was essentially undamaged by beach erosion during hurricane *Hugo*, partly because the setback line had been adhered to.

ESTUARIES

Definition

According to the original definition, which was basically a description of Chesapeake Bay

56

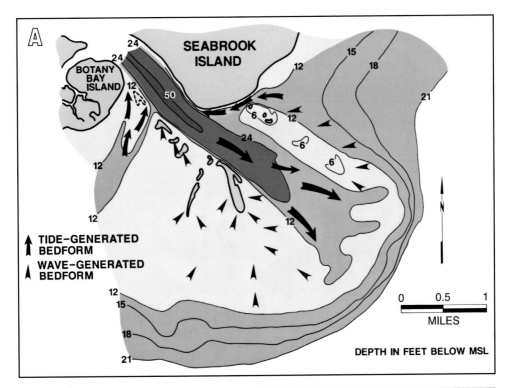

FIGURE 38. Examples of ebb-tidal deltas on the South Carolina coast. (A) North Edisto Inlet. Arrows indicate the dominant orientation of bedforms generated by tidal currents (e.g., megaripples) and wave-generated currents (e.g., ripples). Compare this diagram with the general model given in Figure 37. Note the dominance of landward–directed sand transport in the marginal flood channels and seaward-directed transport in the main ebb channel. Arrows are based on actual field observation through several tidal cycles (Imperato, Sexton, and Hayes, 1988). (B) The ebb-tidal delta of Price Inlet, South Carolina at low tide in the winter of 1975. Compare this ebb-tidal delta with the model in Figure 37. Note the symmetrical ebb-tidal delta, with a clearly defined main ebb channel, terminal lobe, several well-developed swash bars, and two marginal flood channels.

(Pritchard, 1967), an estuary has three defining qualities: 1) a **flooded river valley** that was formed during the lowstand of sea level that culminated about 12-16,000 years ago; 2) a water body with a **substantial freshwater influx**; and 3) a water body **subject to tidal fluctuations**. The second and third criteria are self explanatory, but the first requires some understanding of sea-level fluctuations during the Pleistocene Epoch. As noted earlier, when the sea level was low, streams flowing across the coastal plain of the eastern United States carved valleys up to a hundred feet deep across the coastal plain and the continental shelf. In many cases, this type of erosion was confined to the same valleys during each of the four major drops in sea level. When sea level started to rise at the beginning of the Holocene Epoch about 12-16,000 years ago, those valleys were flooded as the shoreline advanced across the then-exposed continental shelf. When sea level essentially stopped this sudden rise around 4,500 years ago, all but a few of the valleys were flooded with sea water for some distance up the valleys. Where the valleys

were eventually filled with sediment and a bulge of sediment protruded out into the ocean away from the general shoreline trend, the original "estuaries" were converted to what we now call deltas. On the east coast of the United States, this has occurred at only three places – the combined Santee and Pee Dee Rivers in South Carolina, the Savannah River at the South Carolina/Georgia border, and the Altamaha River in Georgia, all three of which are located within the Georgia Bight. Therefore, only the major piedmont rivers (Figure 7) have formed modern deltas in the Georgia Bight.

Characteristics

A key factor in defining the characteristics of an estuary is the way in which the salt and fresh water mix. As shown in Figure 39, a highly stratified estuary with a well-defined **salt wedge** occurs where a significant freshwater stream enters a flooded valley with a small tidal range, with the more dense salt water hugging the bottom of the

TYPES OF ESTUARIES

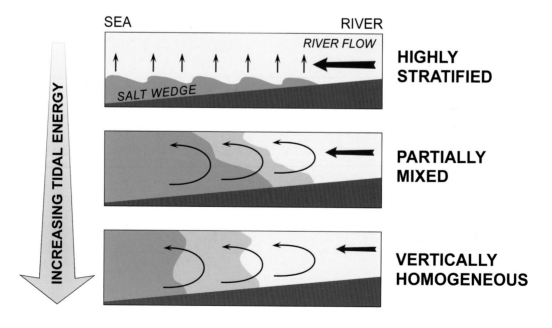

FIGURE 39. Effect of tides on the mixing of salt and fresh water in estuaries (modified after Biggs, 1978).

channel during a rising tide (a highly stratified estuary; see upper diagram in Figure 39). Mixing of the salt and fresh water is enhanced by an increase in tidal energy; therefore, with a larger tidal range, the two water masses become partially mixed (as indicated by the arrows in the middle diagram in Figure 39). With an even larger tidal range, the water within the estuary may become vertically homogeneous from top to bottom, completely fresh top to bottom at the head of the estuary and pure salt water from top to bottom at the entrance (lower diagram in Figure 39). Because of the significant tides, estuaries in South Carolina are usually of the partially mixed type.

Figure 40A shows a plan view model of an estuary of the type found in South Carolina, and Figure 40B shows a cross-section illustrating the circulation within these partially mixed estuaries. We consider the most landward boundary of an estuary to be the place where the tide "stops going up and down," the so-called "head of tides." In the upper portions of the estuary, the primary hydrodynamic process is the river flow, whereas at the entrance, waves and tides dominate. As far as sediment sources are concerned, the sediments in the middle and upper portion of the estuary are provided by the incoming river, but in some estuaries, a considerable amount of sediment comes into the entrance of the estuary from offshore under the influence of tidal and wave-generated currents.

Considering the different components of the system as illustrated by the plan view map in Figure 40A, the upper estuary is a zone of relatively diminished tidal flow, and the river flow is still strong in a seaward direction. In the middle of the estuary, tidal flow is increased, the flood tide reverses the river currents when the tide rises, and a complex two-way, two-layer circulation occurs, creating a zone of mixing of the two water masses. In that zone of mixing, demonstrated in the cross-section in Figure 40B, an important process known as **flocculation** of the clay particles in suspension

occurs. Flocculation takes place because the abundant clay particles in suspension brought in by the river flow slow down their seaward movement at the point of mixing as a result of the reversing of tidal flow directions twice a day. This allows an increase in the number of clay particles in the water column to the extent that they begin to collide with one another. These colliding particles tend to adhere together under the influence of the increased concentration of cations, such as $Na+$, $Ca++$ and $Mg++$, derived from the sea water. As a result, the clay particles, which may have diameters as little as one micron, cluster into groups (flocs) that may have diameters measured in hundreds of microns. As a result, many of the flocs sink to the bottom at slack tide (Figure 40B). Once on the bottom, some of the flocs are resuspended by ebb- and flood-tidal currents. The net result is the formation of a zone in the middle reaches of the estuary that contains abundant fine-grained sediment in suspension known as the **zone of the turbidity maximum.** This zone of turbidity can migrate up and down the estuary over time, moving downstream during periods of flood on the river and upstream during low-water periods. A side effect of this process is the creation of a "mud zone" on the tidal flats and in the tidal channels in the middle of the estuary.

At the estuary entrance, conditions take on an entirely different aspect in that:

- Tidal currents are more dominant than river currents with common separation of ebb-dominated and flood-dominated channels.
- The waters are more mixed, with diminishing estuarine stratification.
- Wave activity becomes a more important dynamic process.
- Moving of sediments along the bottom by wave- and tide-generated currents causes the tidal flats and channel bottoms to become much more sandy than those in the central portions of the estuary in the "mud zone."

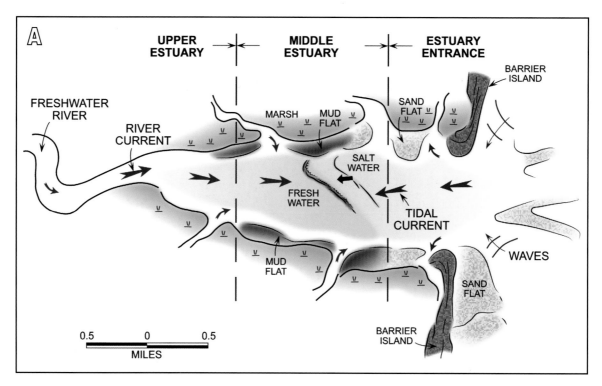

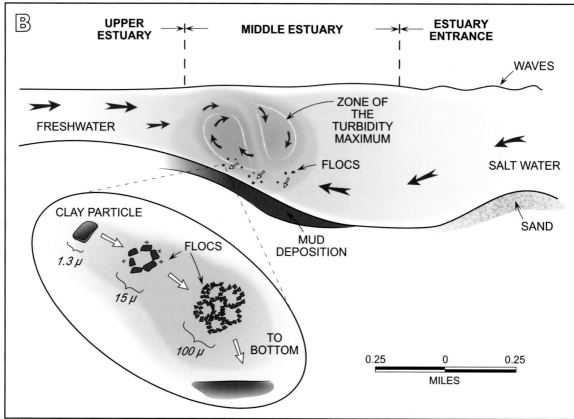

FIGURE 40. Estuaries. (A) Plan view of a typical estuary in South Carolina. (B) Cross-section of a typical estuary showing predominant processes and circulation patterns. Especially noteworthy is the formation of a *zone of the turbidity maximum* where the clay particles (1.3 micron or *μ* in size) brought in by the river undergo a process called *flocculation*.

Estuaries in South Carolina

From north to south, the most prominent estuaries in South Carolina are Winyah Bay, with its associated river systems, including the Waccamaw and Pee Dee Rivers; the three river systems that empty into Charleston Harbor (Wando, Cooper, and Ashley Rivers); Stono estuary; North Edisto estuary; St. Helena Sound, sometimes referred to as the ACE Basin, with its three river systems (Ashepoo, Combahee, and Edisto Rivers); and Port Royal Sound. Because of the limited amount of fresh water that flows into the Stono and North Edisto systems, one could question whether or not they fit the ideal definition of an estuary. This is a term that is much-abused in the literature, but we will not quibble about it here.

DELTAS

As noted in the previous discussion, an irregular progradation of the shoreline at the mouth of a sediment-laden stream is normally referred to as a delta. In their classic summary paper on deltas, Coleman and Wright (1975) discussed over 50 parameters that have an impact on river delta morphology. Factors such as characteristics of the drainage basin, river slope, and coastal climate were acknowledged. Most present-day workers, however, simplify matters by focusing on three basic factors – sediment supply, wave energy, and tidal current energy – in their attempts to classify deltas. Galloway (1975) placed this concept on a ternary diagram, with sediment supply, wave-energy flux (i.e., essentially how big the waves are), and tidal-energy flux (usually determined by tidal range) comprising the three end members. We have superimposed tidal range on Galloway's diagram in Figure 41.

Just considering the end member types, at the top of the triangle in Figure 41 are **river-dominated deltas**, in which a strong river supply of sediment overwhelms the dynamic processes at the shoreline

and a lobe of sediment, such as the birdfoot of the Mississippi River Delta, protrudes out into the sea (see Figure 42A). On wave-dominated coasts, where the rivers have enough sediment to build a bulge out into the ocean, the waves rework the sand into a smooth outer margin composed of prograding beach ridges, such as is illustrated in Figure 42B, a plan view sketch of a **wave-dominated delta**. On tide-dominated coasts, the **tide-dominated**

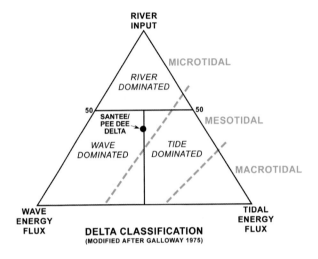

FIGURE 41. Classification of river deltas on a ternary diagram, with three factors controlling the delta morphology: 1) river input (amount of sediment delivered to the coast by the river); 2) wave energy flux (how big the waves are); and 3) tidal energy flux (controlled primarily by the tidal range). We have superimposed tidal range on this diagram, which was originally devised by Galloway (1975). The fields within which the three principle types of deltas – river-dominated, wave-dominated, and tide-dominated – would plot are shown. These three delta types are illustrated in Figure 42. Note the approximate location of the Santee/Pee Dee Delta on the diagram, which we refer to as a mixed-energy delta.

deltas consist of a series of funnel-shaped water bodies (estuaries?) at multiple river mouths with a number of shore-perpendicular, tidal sand ridges that extend offshore of the river mouths (see Figure 42C).

There is only one river delta complex located within the confines of the South Carolina coast, the Santee/Pee Dee delta, which will be discussed

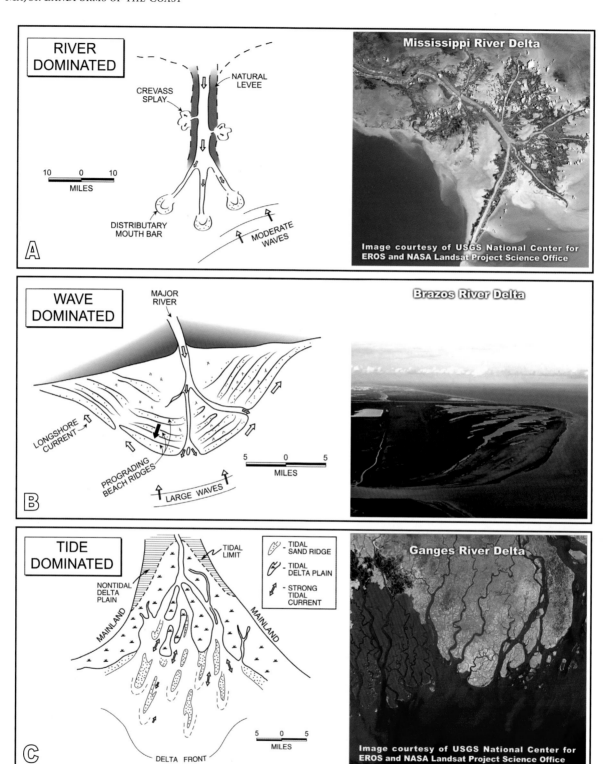

FIGURE 42. River delta models. (A) Schematic sketch in plan view of a river-dominated delta. The main lobe of the delta projects far beyond the shoreline of the adjoining coast. During floods, channels (sometimes referred to as crevasses) may be cut across the natural levees and crevasse splays (minor delta-like lobes) are deposited in adjoining waters. (B) Wave-dominated delta. Note prominence of prograded beach ridges throughout the delta plain. (C) Tide-dominated delta. Note the presence of tidal sand ridges at the offshore entrances to the numerous estuarine complexes along the margin of the delta.

in some detail Section II. We classify the Santee/ Pee Dee delta as a **mixed-energy delta**, plotting it halfway between the tidal- and wave-energy poles on ternary diagram in Figure 41. We do so because marine processes mold the outer margins of the delta, with both waves and tides playing an important role, and the sediment load of the two river systems, though the largest in the Georgia Bight, is modest compared to other rivers around the globe.

SAND BEACHES

A River of Sand

A beach is an accumulation of unconsolidated sediment transported and molded into characteristic forms by *wave-generated* water motion. The landward limit of the beach is the highest level reached by average storm waves, exclusive of catastrophic storm surges, and the seaward limit is the lowest level of the tide. In South Carolina, the beach's landward limit typically is a line of wind-blown sand dunes, called the foredune ridge, or in some relatively rare areas, a man-made structure.

Beaches will form on any shoreline where clastic particles somewhat resistant to wave abrasion are available and there is a site for sediment accumulation protected from extreme daily erosive wave action, such as wave reflection off rocky cliffs. On the South Carolina coast, those clastic particles are mostly sand-sized fragments of rocks composed predominantly of quartz. In a few places, the clastic particles are shells and shell fragments, which may range up to gravel in size.

Most beaches in the state are a virtual "river of sand" with the sand being in constant motion up and down the beach face and along shore. We call the zone within which the sand moves back and forth the **zone of dynamic change** (ZODC), which is illustrated in Figure 43. Part of the wind-blown dunes that form back of the beach above the high-tide line, called foredunes, act as a storage

area for sand between storms. Thus, the landward limit of the zone of dynamic change is the point at which the foredunes are eroded during "normal" storms (not hurricanes). Accordingly, during these "normal" storms, the sand formerly deposited on the intertidal beach, as well as up to several feet of the foredunes, is usually transported into the subtidal area, which, in some areas, contains an offshore bar (see Figure 43; the red zone represents the portion of the beach/dune sand that potentially could be eroded during a typical, moderate storm). The seaward limit of the zone is what engineers call the **closure point**, the most seaward distance for sand to move in an offshore direction under relatively *normal adverse conditions* (not major hurricanes!!). As indicated in Figure 43, all of the sand grains in motion do not stay within the zone of dynamic change at a specific geographic site on the beach for an infinite amount of time, but most likely will eventually "escape" from the zone by one of three primary types of "leakage:"

1) **Alongshore** out of the zone, under the influence of longshore sediment transport by wave-generated currents;
2) **Offshore** beyond the point of closure during major storms, such as hurricanes; and
3) **Onshore** by wind action, beyond recovery by offshore winds or by erosion during storms.

Another way sand grains may "escape" from the "river of sand" for a long period of time is for the shoreline to build out seaward to the extent that more lines of foredunes develop until they are located far enough landward to be out of the reach of normal storm waves.

The Beach Profile

When coastal geologists or engineers study beaches to determine such issues as their potential for erosion, they usually measure a topographic profile across the beach surface using a variety of

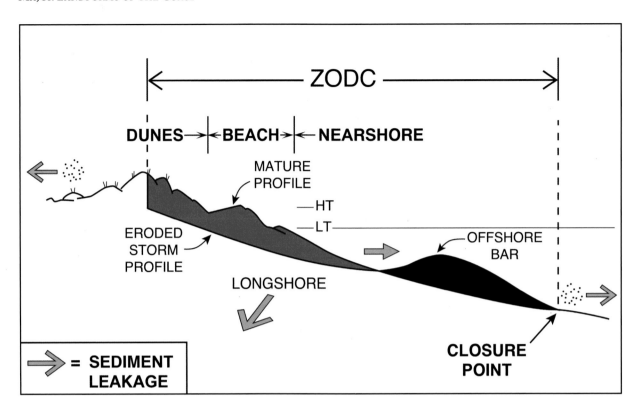

FIGURE 43. Zone of dynamic change (ZODC). The area in red depicts sand that may be eroded from the beach area and deposited offshore during storms. Much of this sand is deposited on the offshore bar (colored black). In between storms, much of the eroded sand is returned to the beach/dune area. The yellow arrows indicate the three mechanisms by which the sand might be lost from this particular ZODC: 1) carried offshore beyond the closure point during large storms, such as hurricanes; 2) blown landward out of the foredune area during periods of high wind activity; and 3) transported alongshore out of the ZODC by beach drift and longshore currents as illustrated in Figure 22.

surveying techniques. This is usually repeated over a period of months or years in order to determine the cycle of erosion and/or deposition on that particular beach. Such studies have been carried out along many of the coastlines of the world, including the coast of South Carolina (by Hayes and his students and numerous others). At the present time, a program to make periodic measurements of the beach profile at 400 sites located mostly on the developed portion of the coast is being carried out by the South Carolina Department of Health and Environmental Control (DHEC), Office of Ocean and Coastal Resource Management (OCRM) with funds provided by the U.S. Geological Survey. "State of the Beaches" reports by this group, which include summaries of the changes of the profiles at the different locations, can be accessed at http://

www.scdhec.net/environment/ocrm/pubs/docs/reports.htm#beaches.

For purposes of discussion, it is necessary that the different morphological components of the intertidal beach be defined. In the early days of research on the California coast, the profile shown in Figure 44A was proposed as typical for that area. Under that scheme, the beach was divided into two segments: 1) Backshore – the landward sloping surface of a broad depositional feature called a **berm**; and 2) Foreshore – the seaward face of the berm also alluded to as the **beach face** in that scheme. The subtidal area was divided into the surf zone, breaker zone, and offshore. This type of profile most commonly forms on a windward shore subject to ocean swell and a relatively large average wave height, such as occurs on the California coast.

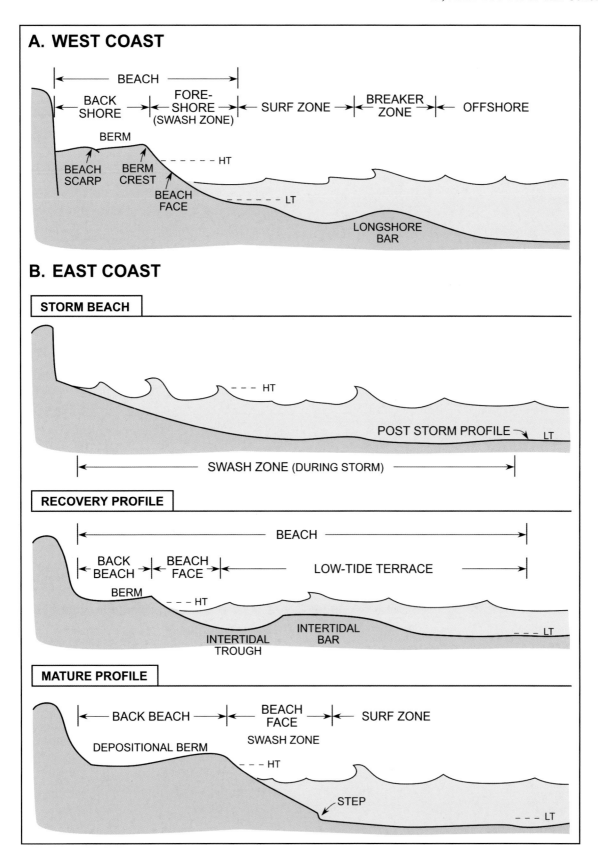

FIGURE 44. Representative beach profiles from the west (A) and east (B) coasts of the United States.

Although there have been some efforts to force fit this California nomenclature to the sand beaches of the east coast of the United States, we prefer to use the nomenclature in Figure 44B. The east coast is a leeward shore dominated by storm waves and the average wave height is considerably less than that experienced in California. Also, the influence of the relatively large tidal ranges on parts of the east coast is an important consideration. The upper profile shown in Figure 44B is the flat and featureless surface left after a storm. The middle profile is the one that most commonly occurs on the South Carolina coast during intervals between storms. It consists of three segments: 1) **back beach**, or the landward sloping portion of the depositional berm; 2) the **beach face**, the zone where the waves swash back and forth at high tide (seaward face of depositonal berm); and 3) the **low-tide terrace,** which commonly contains an **intertidal bar** and an **intertidal trough**. In situations where the depositional cycle is complete, or where the beach borders a tidal inlet, the configuration shown in the bottom of Figure 44B may be present, which is similar to the typical California profile.

A key aspect of any beach profile is the slope of the beach face. Beach research as early as during World War II focused on this issue, due to its relevance to beach landings by amphibious watercraft. Those early researchers found that the slope of the beach face in any given area will attain an equilibrium profile (which includes a near constant slope of the beach face) under what they referred to as "steady" conditions, that is, relatively unchanging, day-to-day wave characteristics. During storms, sand beaches are commonly eroded down to a flat surface that essentially parallels the water table in the beach with sand typically being transported offshore (see upper profile in Figure 44B). When sediment returns to the beach after the storm, the depositional profile attained by the beach face is steeper than the beach was immediately following the storm. Komar (1976, p. 303) explained why:

"Due to water percolation into the beach face and frictional drag on the swash, the return backwash tends to be weaker than the forward uprush. This moves sediment onshore until a slope is built up in which gravity supports the backwash and offshore sand transport."

There's that law of gravity again!! Several field studies and wave tank experiments have shown that coarser-grained beaches, especially those composed of gravel, have steeper slopes than those composed of finer-grained sediments. Coarser-grained beaches allow greater percolation of the water brought to the beach face in the wave uprush than finer-grained beaches, thus reducing the strength of the return of the backwash. Consequently, a steeper beach face is required to maintain an equilibrium profile on coarse-grained beaches (gravity supported transport). On sandy beaches of comparable grain sizes, those beaches consistently exposed to large waves tend to be flatter than sheltered beaches exposed to smaller waves, because larger waves produce stronger backwash, which tends to flatten the profile.

On the South Carolina coast, most sand beaches are composed of fine-grained sand, consequently they tend to be quite wide and flat. An exception occurs where the beach flanks a major tidal inlet channel, where the slope of the beach face is an extension of the slope of the channel itself. The steepest beaches in the state occur where the beach is composed of pure, gravel-sized material, mostly oyster shells. These steep shell beaches occur in three types of settings in South Carolina:

1) Along the Intracoastal Waterway, where boat wakes erode oyster mounds in the adjacent marshes and tidal flats with the eroded shells being deposited in beaches along the channel edge;

2) On the windward sides of some of the more open estuaries, where at high tide the waves erode the oyster mounds in the marshes and

tidal flats and a ridge of coarse shells (storm berm) is deposited on the adjacent, eroding marsh surface (e.g., the channel of the Edisto River in the lower reaches of St. Helena Sound); and

3) On some of the outer, landward-migrating barrier islands, where the marsh and tidal channels exposed to wave action in the intertidal zone contain oyster mounds that are eroded as the island migrates landward (e.g., islands on the outer shoreline of Cape Romain).

The Beach Cycle

The concept of the cyclical change of beaches from a flat, erosional profile in winter to a wide, depositional berm in summer had been well ingrained in both the popular and scientific literature for many years following research carried out in the 1940s. The concept originated from detailed studies on the coast of southern California (e.g., Bascom, 1954). The idea was that, generally speaking, erosional storm waves are more common in the winter, and flatter, depositional swell waves are more common in the summer on the California coast – hence the terms summer and winter beaches. This is an idea that has not withstood the test of time. In 1969, Hayes and colleague Cy Galvin observed a striking contrast between beach cycles on the U.S. east coast and those on the west coast. Eroded winter profiles and strongly depositional summer profiles are not a dominant theme on northern Atlantic beaches, particularly in the New England area. The beach cycle in that area is controlled by the passage of individual storms. This also appears to reflect, at least to some degree, less severe wave climate and relatively smaller seasonal changes on the northern Atlantic coast. Mean wave heights and periods are 19 inches and 6.9 seconds along the northern Atlantic coast and 28 inches and 12.5 seconds on the California coast. These differences in wave climate result from differences in location on western and eastern edges of oceans

with predominantly westerly winds, winds blowing predominantly onshore on the California coast and offshore on the east coast. In the 2002 Coastal Engineering Manual, published by the U.S. Army Corps of Engineers, the following statement appears: *"the seasonal onshore-offshore exchange of beach and nearshore sediments … is now known to apply best to swell-dominated coasts, such as the US west coast, where wave climate changes seasonally."* Therefore, the general consensus today is that the terms "winter beach" and "summer beach" should not be used with reference to erosional and depositional cycles on beaches. Hayes and associates in their work in New England used the terms early post-storm profile and late accretional profile (neither of which is restricted to any particular season of the year), and Paul Komar, working on the Oregon coast, suggested storm profile and swell profile.

Based on six years of research, Hayes and his students at the University of Massachusetts found that on the northern New England coast, northeasterly storms play the dominant role in the generation of cycles of erosion and deposition of the sand beaches in that area (Hayes and Boothroyd, 1969). Observations at 40 beach profile stations over that six-year period revealed the following stages of beach morphology relative to storm occurrences:

1) **Early post storm** (up to 3 or 4 days after the storm) – Beach is flat to concave upward and the beach surface is generally smooth and composed of medium- to fine-grained sand. Severest storms leave erosional scarps along the dunes (see top diagram in Figure 44B).

2) **Early accretional, or constructional** (2 days to 6 weeks after storm) – Small berms at the high-tide line and intertidal bars and troughs are quick to form as waves bring the sand back onto the beach (see middle diagram in Figure 44B).

3) **Late accretional or maturity** (6 weeks or more after storm) – Landward-migrating intertidal bars weld onto the backbeach to form broad,

convex berms (see lower diagram in Figure 44B). On some beaches, it takes a long time for welding to occur and large intertidal bars (up to 3 feet in height) lie between the back beach and the low-tide line (see example from South Carolina coast in Figure 45).

Our work on the South Carolina coast shows that the South Carolina sand beaches follow a cycle similar to those in northern New England, particularly those close to tidal inlets (Hayes et. al., 1975; Stephen et al., 1975). In mid-barrier areas away from inlets, however, the beaches are usually quite flat or contain intertidal bars that migrate landward very slowly. A completed cycle like the one illustrated in the lower diagram in Figure 44B is rare indeed in mid-barrier areas, but they can be seen on occasion in the near vicinity of tidal inlets.

Physical Sedimentary Features and Biological Components

Because of the relatively large tides and fine-grained sand on the beaches in South Carolina, when you visit the beach at low tide you will be able to walk across a wide intertidal zone that contains an abundance of both physical and biological features that will no doubt arouse your curiosity. Although geology has long been considered to be primarily a field science, the greatest breakthrough in the understanding of the smaller-scale physical sedimentary features you will be walking on at the beach was the result of laboratory flume studies carried out by hydraulic engineers (e.g., Simons and Richardson, 1962). In flowing water, sand grains start to move at velocities of about 10 inches/second. These workers demonstrated that the sediment

FIGURE 45. Low-tide photograph taken in May 1974 showing a large intertidal bar on the north end of Kiawah Island, South Carolina. Antidunes in foreground were formed when shallow sheets of water pushed forward by breaking waves (at high tide) flowed rapidly down the landward side of the bar.

bed of a laboratory flume (box-like trough with a sand bottom and a mechanism to generate water flow across the sand bed) could be made to pass through the following sequence of features (called **bedforms**) by simply increasing the velocity of the water flowing across the sand bed or changing the depth of the water:

> *flat bed → small ripples → large ripples (megaripples) → washed-out megaripples → flat bed → antidunes*

While doing this work, they came up with an empirical relationship that dictates which of these different bedforms are active at any given time. It is called the **Froude number**, which equals:

$$\text{Froude number} = \frac{V}{\sqrt{gD}}$$

V = velocity
 g = gravitational constant
 D = water depth

Therefore, as the water velocity increases, the Froude number also increases, and as depth increases, the Froude number decreases.

The first part of the sequence up to the washed-out megaripples occurs when the value of the Froude number is <1.0. This part of the sequence is called the **lower flow regime**. By increasing the Froude number to values >1.0, by either increasing the velocity of flow or decreasing the water depth (see formula), the sand bed goes into the **upper flow regime**, where either flat beds or antidunes are formed.

The lower flow regime bedforms are asymmetrical, with the sand grains moving in discreet steps by rolling or bouncing up the gently sloping upcurrent side of the bedform and sliding down the steeply dipping surface on the

downcurrent side (called the slip face). Thus, these forms move along in the direction the current is flowing, sometimes several feet during a single tidal cycle. On a beach, it is rare to see any bedforms larger than **ripples** (spacing between crests less than 2 feet) and they usually occur in the intertidal troughs on the landward side of the intertidal bars, where the water depths are greater than elsewhere on the beach (see example in Figure 46A). **Megaripples** (spacing between crests 2-18 feet) are seen more commonly in tidal channels, where the water is deeper and the currents are stronger (see Figures 46B and 46C). The mode of sediment transport over these types of bedforms is illustrated in Figure 47. Larger bedforms called **sand waves** (spacing between crests greater than 18 feet) do occur on this coastline, but they are mostly found in deep tidal channels.

As you walk across the beach and tidal sand flats at the coast, you will notice some dramatic changes in the nature of the lower flow regime bedforms (at the scale of ripples), even though the changes in the topography are quite subtle. As the diagram in Figure 48 shows, there is a progression of the three-dimensional shapes from straight-crested to undulatory ripples at conditions of relatively low flow strength. As flow strength increases, the ripples take on a cuspate shape. This shape is very common in the intertidal troughs on the beach. As the flow approaches the upper flow regime (Froude number >1.0), the ripples convert to a planed-off rhomboid shape just before the bed goes flat. The photographs in Figure 49 illustrate these bedform types. You will undoubtedly see the pattern illustrated in Figure 48 repeated time and time again during your walks over the intertidal bars or near the edge of tidal inlets during low tide. The pattern will almost always be present where the high-tide waters were flowing down very subtle slopes on the sand bars.

The most common bedform type under upper flow regime conditions is the **flat bed**, over which the flow of the sand grains is more-or-less continuous, streaming in long streaks across the

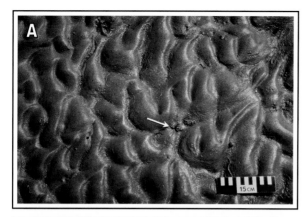

FIGURE 46. Ripples and megaripples. (A) Ripples exposed at low tide on a tidal river point bar in the South Edisto Estuary. Arrow points to exposed ghost-shrimp burrow. (B) Megaripples exposed at low tide on a sand flat at Hampton Harbor Inlet, New Hampshire. View from bridge. Arrow points to megaripple shown in C. The ebb current that created these megaripples was flowing from the lower left to the upper right. (C) Ground view of the megaripple shown in photograph B. Note the high angle of dip of the beds created by the migration of the bedform by the mechanism illustrated in Figure 47. The machete is about 2.5 feet long. Ebb current that created this bedform was flowing from left to right. Photographs in B and C were taken in November 1965.

sediment surface. This is the most common type of sand surface occurring on sand beaches, where water depths are typically small and wave-generated water flow is generally rapid. Thus the Froude number is typically >1.0 (see formula again). Under conditions where the velocity is more rapid than normal, for example, when the water flows down a subtle slope [e.g., over the crest of an intertidal bar (Figure 45) or down a more steeply sloping high beach face], a feature called an **antidune** is formed. Under this condition, the sand surface develops a sinusoidal shape and a relatively thin layer of water flows in phase across the top of these features, which are linear and parallel with a typical spacing of about 15-20 inches. As shown in Figure 50A, the water flowing down the slope on the downcurrent side of the sinusoidal antidune picks up sand grains from that relatively steep slope only to drop them out again as the water flows up the slope on the upcurrent side of the next antidune in line. A slice through a preserved antidune on a South Carolina beach, shown in Figure 50B, illustrates the deposited sand on the steep upcurrent side of the antidune. At first glance, as you watch antidunes in motion, you experience an optical illusion that the sand grains are moving upstream against the current. Actually, you are viewing the individual sinusoidal forms of the antidunes that are moving against the current, because of erosion on the downcurrent side and deposition on the upcurrent side. Although they move in a series of stops and starts, the sand grains always move in the direction the water is flowing.

If you look carefully, you will almost always see antidunes forming on the South Carolina beaches in one of two places:

1) Where water draining from intertidal trough outlets cut through and across the intertidal bar located just seaward of the trough; and

2) Anywhere along the upper or lower part of the beach where the backwash flow gains velocity as it moves back down the beach slope. A photograph of this process is shown in Figure

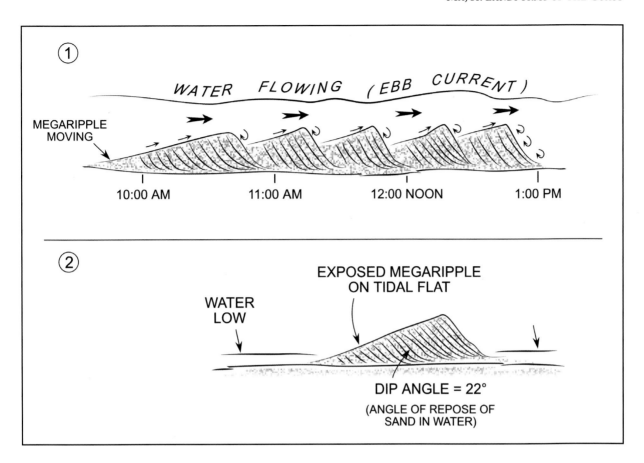

FIGURE 47. Mechanism for the movement of megaripples. Compare the implied slip faces of the bedform as it moves along (in the sketch) with the ones exposed in the excavation in the photograph in Figure 46C.

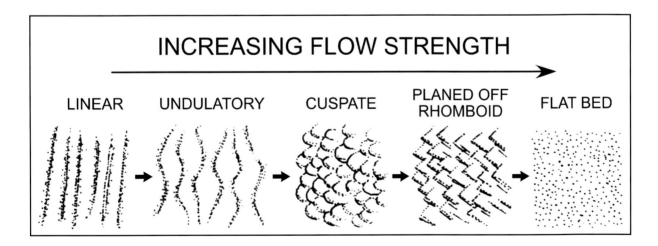

FIGURE 48. Changes in linearity of crests and 3-dimensional shape of ripples under conditions of increasing flow strength. Based on data of Allen (1968), Boothroyd (1969), and numerous field observations by our group on the intertidal bars and sand flats on the South Carolina coast. In all cases, the water was flowing from left to right.

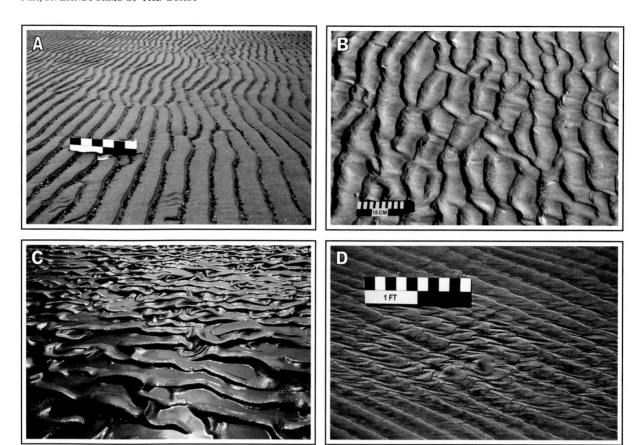

FIGURE 49. Photographs of the ripple types illustrated in Figure 48. (A) Linear ripples. Scale is one-foot. The ebb current that created these ripples was flowing from left to right. (B) Ripples with undulatory crests, formed by an ebb current flowing from left to right. Scale is 15 centimeters (~ 6 inches). (C) Cuspate ripples, formed by an ebb current flowing from the upper right to the lower left. (D) Small rhomboid ripples. Flow is from left to right. Scale is one-foot. This ripple form is the last to develop before the bed passes to the flat-bed stage in the transitional zone between lower and upper flow regime.

51A. An example of antidunes preserved on the upper beach face at Kiawah Island is shown in Figures 51B.

It has been our experience that students walking in the swash zone get pretty excited when they feel the antidunes move by past their feet. Sand moving uphill!! Not exactly, but close.

The shallow, rapid flow of water over the antidunes tends to sort the sediment by composition, moving out the larger and lighter quartz and feldspar grains and leaving behind more dense and smaller grains, such as hornblende [$Ca_2(Mg,Fe,Al)_5(Al,Si)_8O_{22}(OH)_2$], magnetite ($Fe_2O_4$), and ilmenite ($FeTiO_3$). These more dense minerals, called **heavy minerals**, are darker in color than the lighter minerals, hence they leave linear traces that mark the position of the antidunes in which they were deposited (see Figure 51B). Another place where you will see relatively thick deposits of heavy minerals is at the base of an eroding dune scarp at the back of the beach, where the wind has carried the lighter grains away and left the heavier ones behind. Occasionally, the high part of the beach face is covered with a layer of them, usually after a storm, as shown in Figure 52A. A variety of minerals occur in these black layers (Figure 52B). One of the most common minerals in these black layers, ilmenite, is rich in titanium, and deposits similar to the ones you will see on the beach are also abundant in some of the Pleistocene

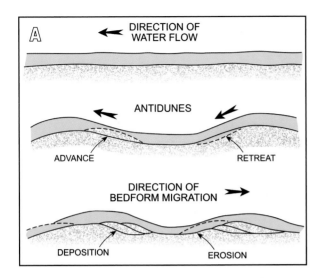

FIGURE 50. Antidunes. (A) Formation and growth. As the shallow sheet of water flows across the linear form of the antidune, water flowing down the downcurrent side of the antidune tends to erode sand from that slope. As it flows up the slope of the next antidune down current, sand is deposited on that slope, causing the feature to migrate upcurrent. This bedform is called an antidune, because it moves in the opposite direction to the current. Lower flow regime ripples, on the other hand, move in the same direction as the current (compare this figure with the diagram in Figure 47). However, the sand grains themselves move downstream in both situations. (B) Trench through antidunes on the beach on Seabrook Island, South Carolina (photograph taken at low tide in June, 1975). Water was flowing from right to left when the antidunes were forming. Compare the internal sand layers in the preserved antidune with the sketch in diagram A. Scale is one foot.

beach ridges landward of the coast, particularly in Georgia and Florida. There has been some mining of these Pleistocene heavy-mineral deposits, but environmental concerns are an issue.

Two of the most conspicuous biologic features on the South Carolina beach, besides the feeding birds and the occasional stranded seashell, are burrows by ghost crabs (*Ocypode quadrata*) just landward of the high-tide line, and the abundant burrows of the ghost shrimp (*Callianassa major*) near the low-tide line. These burrows will be illustrated and reviewed in more detail in Section II, under the discussion of Beachwalker Park on Kiawah Island, one of the best places on the coast to see these features.

As mentioned earlier, the most common beach profile in the mid-barrier region contains a modest-sized high berm and a well-developed intertidal bar, as is clearly shown by the oblique aerial photograph of the middle of Kiawah Island in Figure 53. The illustration in Figure 54A shows the distribution of some of the physical and biological features typical of this type of beach:

- Antidunes on the high-tide and low-tide swash zone area;
- The low angle of dip of the beds on the swash zones and the high-angle of dip of the intertidal bar's slip face;
- Ghost shrimp burrows near the low-tide line; and
- Ghost crab burrows above the high-tide line.

Figure 54B illustrates the typical late-stage constructional profile (welded berm) commonly found near tidal inlets, as well as the distribution of the physical and biological features found there, including:

- The surface of the beach just after the last storm (dashed line);
- Antidunes on the berm top and beach face;
- The low angle of dip of the beds formed during

FIGURE 51. More antidunes. (A) In the lower left of the image, antidunes are forming in the backwash of a wave. Further up the beach, preserved antidunes are in evidence. This sunset picture was taken on the beach at Seabrook Island, South Carolina in June 1975. (B) Antidunes preserved on the upper beach face on the northern end of Kiawah Island, South Carolina. The return flow that created the antidunes also sorted the sand grains by composition, leaving behind bands of black heavy minerals dominated by magnetite and ilmenite. Scale is 15 cm (~ 6 inches). Photograph taken circa 1980.

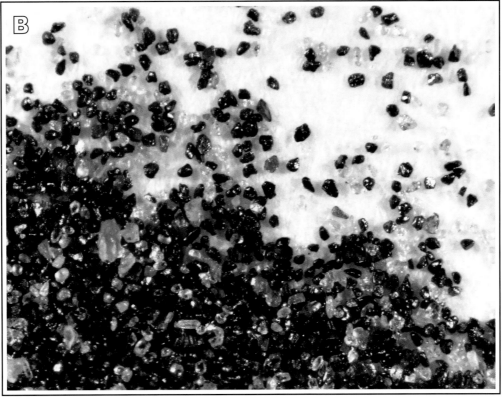

FIGURE 52. Heavy minerals. (A) Beach trench in heavy mineral zone at north end of Kiawah Island. Internal structures in trench indicate that a depositional berm had been truncated by an erosional episode that left a 1-3 inch thick deposit of heavy minerals on the truncation surface. Photograph taken in May 1978. (B) A 3.5 mm field of view of a sand sample from a heavy mineral layer on a South Carolina beach. The black minerals are dominated by magnetite (Fe_2O_4) and the clear grains are mostly quartz (SiO_2). Courtesy of Ray Torres.

FIGURE 53. Intertidal bars at low tide, central Kiawah Island (infrared photograph taken circa 1978). Waves were slowly moving the bars onshore, with the potential of their welding onto the beachface, forming a wide beach. This process, interrupted at times by major storms, has taken place numerous times over the years.

the later stages of development of the berm and the high-angle of dip of the slip face of the intertidal bar that first welded to the beach;

- Ghost shrimp burrows near the low-tide line; and
- Ghost crab burrows above the high-tide line.

COASTAL DUNES

Extensive fields of barren coastal dunes that can be reached by automobile are only present (in 2007) at one location on the South Carolina coast, the outer shore of Seabrook Island. Those dunes are

present at that location because of a sudden major addition of sand to the beach due to an artificial manipulation of the tidal inlet just to the north (discussed in detail in Section II), and because the natural growth of beach vegetation needed to trap the sand has been hindered by beach manicuring. Despite the paucity of barren dunes, all of the prograding-type barrier islands in South Carolina have well-defined lines of foredunes adjacent to the backbeach. Vegetation plays a very important role in the formation of these dune lines, which, over time, continue to build out seaward such that many lines of ancient foredunes perched on top of old beach sediments (called **beach ridges**) cover the surface of those islands that are in the prograding mode.

Figure 55 shows the general model for the evolution of such prograding beach ridges on recurved spits, such as the one shown in Figure 23. A similar, but slower, process occurs in the middle of the prograding barrier islands. This model is based on observing numerous such dune ridges evolve on the prograding barrier islands of South Carolina since 1972. Similar observations have been reported for the South Carolina dunes (Imperato et al., 1983), Georgia foredunes (Oertel and Larsen, 1976) and on Nauset spit, Massachusetts (Hine, 1979). The process involves the following steps:

1) An intertidal bar welds to the backbeach and is molded into a depositional berm by wave action, as was described in the previous section on sand beaches. The crest and top of the berm are built above normal high tides.

2) During an ensuing spring tide, a line of vegetative matter called **wrack**, which is usually composed of stems of the low marsh plant smooth cordgrass (*Spartina alterniflora*) accumulates on top of the berm. The wrack will later serve as a seed bed for beach grasses.

3) Small wind-blown sand dunes, called **wind-shadow dunes**, will rapidly accumulate in the lee of the wrack as soon as the normal land-

A. MID-BARRIER CONSTRUCTIONAL PROFILE

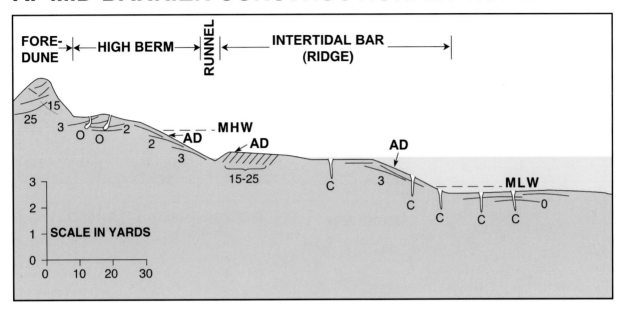

B. NEAR-INLET CONSTRUCTIONAL PROFILE

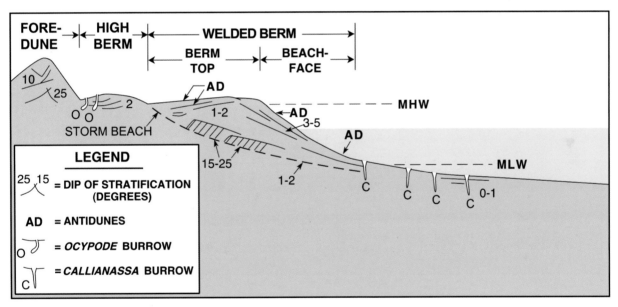

FIGURE 54. Representative morphology and sedimentary features occurring on the beaches of the prograding, mixed-energy barrier islands of South Carolina. Based on inspection of hundreds of beach trenches. (A) Features found on a typical mid-barrier beach in a constructional mode (modified after Hoyt and Weimer, 1963). (B) Features found on beaches near tidal inlets in a late constructional mode.

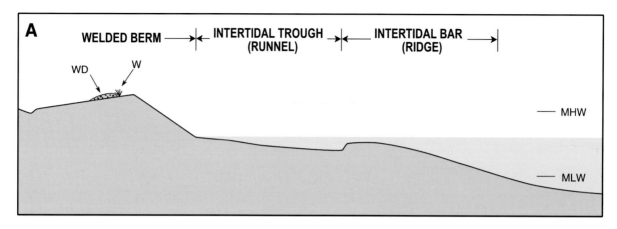

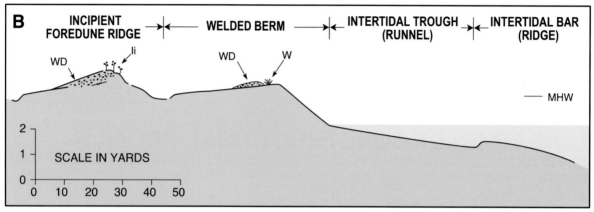

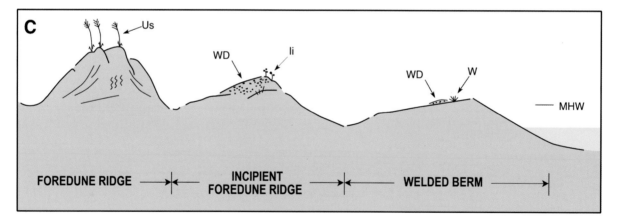

LEGEND

W = Wrack
WD = Wind-shadow Dune
Ii = *Iva imbricata*
Us = *Uniola sp.*

FIGURE 55. Model for evolution of foredune ridges on a prograding, recurved spit in South Carolina (see photograph of one such recurved spit in Figure 23).

directed breezes start to blow. Shortly thereafter, if it is the growing season, salt-tolerant beach grasses, such as sea elder (*Iva imbricata*), grow out of the wrack and become baffles to the wind action as well, creating even larger wind-shadow dunes.

4) A new intertidal bar welds to the beach and a broad berm forms that provides a source for more wind-blown sand to nourish the original dune line, which will eventually achieve heights up to 10s of feet. Sea oats (*Uniola sp.*) populate the tops of these higher dune ridges.

5) The process just described in steps 1-4 takes place on the new berm.

After approximately 15 years, wax myrtle (*Myrica cerifera*) will grow in the swales between the dunes. After approximately 50 years, a climax maritime forest, composed of three dominant tree species – live oak (*Quercus virginiana*), palmetto (*Sabal palmetto*), and loblolly pine (*Pinus taeda*) – will begin to populate the beach ridge (P. Hosier, pers. comm.; Hosier, 1975). These trees and shrubs will be pruned noticeably by salt spray as long as they remain on the front line of the forest. Through time, more beach ridges will accrete, and the original dune ridge will be located well inland of the beach. Figure 28 shows the multiple beach ridges that make up the northeast end of Kiawah Island.

6 BEACH EROSION

INTRODUCTION

Erosion is a common phenomenon on many of the shorelines of the world. In fact, almost no coastal area of the world that has been developed by man is free of problems caused by beach erosion. Various publications have given numbers on beach erosion rates; for example, Dean and Walton (1975) reported an average erosion rate of between 1-3 feet per year for the shoreline of the United States. Another study by the U.S. Army Corps of Engineers (National Shoreline Survey) in 1971 concluded that nearly 43% of the U.S. shoreline is undergoing "significant" erosion.

In a paper published in 1985, Hayes made the following somewhat pompous assertion:

"What becomes evident in this discourse is the conclusion that man himself has caused most of his own beach erosion problems. That being the case, perhaps he can do something to solve them. Lessons learned from the mistakes of the past should be applied in the planning phases of every new development in the coastal zone. The costs involved in preventing erosion by wise planning are orders of magnitude less than those required to solve an erosion problem once the structures are in place."

Now, in reviewing what has been happening in the 22 years since that pronouncement, one would have to conclude that either: 1) not more than a dozen policy makers read it, or 2) if they did, they had a good laugh, ignored it, and went on about their business.

Getting back to the numbers, a later estimate by Bird (1996) concluded that 70% of the world's beaches are now eroding. Therefore, it should come as a surprise to nobody that at least some of the beaches in South Carolina are eroding. Three relatively recent reports made the following assessment:

1) Tim Kana, in a 1988 report sponsored by the South Carolina Sea Grant Consortium, stated that for the 50% of the South Carolina coast that is developed, a third of it is "eroding at more than one foot per year," but that "60 miles of developed oceanfront are stable or eroding at less than one foot per year."

2) According to the 1997 annual DHEC report on the status of the beaches, more than 40% of the shoreline of South Carolina is "stable or increasing" at the present time. That report also states that "about 40% is eroding at less than three feet per year, and about 20% is eroding at a rate of more than three feet per year."

3) The 2006 annual DHEC report on the status of the beaches revealed that beach renourishment projects to combat beach erosion were either under way or being contemplated for beaches in all of the major coastal counties. Several of these projects will be discussed in Section II.

Because nearly half of the South Carolina coastal zone is locked up in wildlife refuges, state parks, and other types of natural areas, where erosion is occurring in those areas it is not generally considered to be a "problem." Notable exceptions to this generalization are at Hunting Island and Edisto Beach State Parks, where recreational beaches used by large numbers of people are under constant threat of erosion. However, most of the "problem" areas are in populated regions, where erosion is a result of several factors, in some cases due to unwise coastal development practices and in others where public works, such as jetty structures, have had inconvenient side effects.

MAJOR CAUSES OF BEACH EROSION

Introduction

This discussion of the causes of beach erosion will be limited to depositional shorelines like the South Carolina coast. The causes for erosion on leading edge and glaciated coasts are complex and an understanding of them is not relevant for the coastal area under consideration in this book. With that given somewhat limited focus, we can say once more that for any given depositional coastal area not unduly influenced by man, it is the interaction of sediment supply and water-level changes that controls beach erosion. In some areas, fluctuations in sediment supply prevail, and in others, rapid changes in water level are most important. The most serious erosion problems occur either where man interferes with sediment supply or where some phenomenon causes an unusually rapid rise in water level (e.g., abrupt lake level rise in Lake

Michigan or around sinking abandoned delta lobes on the Mississippi Delta).

Deficits in Sand Supply

Sediments on South Carolina's beaches are dominated by fine-grained sand, with coarse-grained shells and shell fragments being present in a few areas. Quartz is the most common constituent of the beach sand, because of its relative abundance in the earth's crust, as well as its chemical stability and resistance to abrasion. In tropical regions, the beach sands are frequently dominated by carbonate fragments of different origins and in glaciated regions and along mountainous coasts coarse-grained gravel composed of rock fragments occurs on many beaches. In effect, the waves will build a beach from whatever material is available.

On the South Carolina coast, the sand on the beaches was ultimately derived from erosion of the Appalachian Mountains and brought to the sea by streams. In many cases, much of the sand stopped somewhere along the way to be incorporated into ancient sedimentary rocks or sediments, some of which was eroded and carried to the sea during later geological episodes. Once in the general vicinity of the present shore, the sand on any particular beach could have been derived from one or a combination of the following sources:

1) **Carried directly to the shore by streams** during the last few thousand years. With regard to sand beaches in the vicinity of the outlets of the piedmont, or red-water, rivers (Pee Dee, Santee, or Savannah), some fraction of this sand came all the way down the rivers from the Piedmont or Blue Ridge Mountain region, but some of it could also have been eroded from rocks along the stream ranging in age from Pleistocene all the way back to the Cretaceous (~100 million years ago). There is no evidence that sediment delivered to the shoreline by the Pee Dee and Santee Rivers has been transported

much further south than Cape Romain during the present highstand of sea level (the last 4,500 years or so), a distance of about 20 miles. All of the sediment delivered to the shoreline by coastal plain, or black-water rivers, would have been eroded from local sedimentary deposits and not derived directly from the mountains. None of the coastal plain rivers have built deltas on the open ocean shoreline, thus their contribution to modern beaches is probably small.

2) **Eroded from shorelines of different kinds to the northeast** of the site in question and moved along the shore as part of the longshore sediment transport system into the present site. These more northerly sites could have been older Pleistocene shoreline sediments, active and abandoned delta lobes, eroding barrier islands, and so on.

3) **Transported from the continental shelf** to the present beach. The sands on the shelf were most often placed there as river deltas, lowstand valley fills, or other shoreline deposits at times when sea levels were lower than at present. The landward transport of this sediment to some of the active beaches is made obvious by the composition of the sand and by observations on the beach after storms, but the actual process of landward transport is poorly understood.

4) **A variety of processes at tidal inlets**. Studies by our group of a number of tidal inlets in South Carolina show that erosional/depositional patterns of adjacent beaches are controlled closely by morphological changes of the inlets (e.g., Hayes et al., 1976; FitzGerald and Hayes, 1980). One major way for deposition on a beach to occur is for large individual swash bars on the ebb-tidal deltas of tidal inlets to migrate onshore, as is illustrated in Figures 56A and 56B. As the inlet changes, and the bars shift position, the location of the focused wave energy will change. This allows major shifts in deposition on the beaches on either side of the inlet over time.

5) **Artificially added to the beach** by engineers in what is termed **beach nourishment**, a practice that has been taking place on the South Carolina coast with more and more intensity over the past few years. The sand itself can come from a variety of sources, with dredging of offshore sand sources being one of the most common.

To understand which of the four natural sources is most important at a given site, a sediment budget would have to be calculated in which the sediment concentrations (credits) and losses (debits) are determined and equated to the net gain or loss. This kind of sediment budget has been attempted for the California coast (Bowen and Inman, 1966), but not for the South Carolina coast, although it was the subject of a failed grant application Hayes submitted during his University days. Too bad. However, a partial one involving beach volume changes along the central South Carolina coast was completed and published in 2001 by Kana and Gaudiano.

There are a number of natural processes that could produce a deficit of sand at a particular beach, including:

1) **Switching of the mouth of the main river** in a delta system. This is a common process on river-dominated deltas such as the Mississippi Delta. Even the mouth of the Santee River has switched from time to time, causing at least some minor sand deficits along parts of that delta front.

2) **A variety of processes at tidal inlets**. For example, waves refracting around the ebb-tidal delta create a zone of deposition downdrift of the inlet, thus diminishing the flow of sand further down the beach (Figure 29A). Sand bars affiliated with the ebb-tidal delta also create local zones of erosion by focusing wave energy through wave refraction (see June 1983 diagram in Figure 56B). And, of course, as the inlet migrates, the shoreline downdrift is

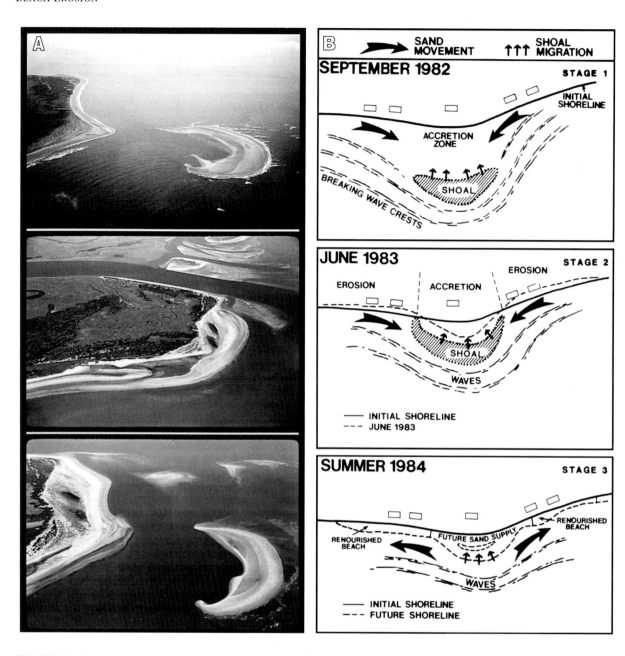

FIGURE 56. Shoreline changes associated with tidal inlets in South Carolina. (A) Large swash bars detaching from the ebb-tidal delta at Stono Inlet and merging with the beach. Upper – Early stage of detachment in 1977. Middle – Attached to shore in 1983. Lower – A new swash bar detached in 1986, which attached to the beach in 1990. From Kana et al. (1999). (B) Illustration of the three stages of bar welding from the same article.

eroded away and the sand on the beach on the downdrift side of the inlet is carried offshore where it is stranded on the ebb-tidal delta for a period of time, which may be years in duration. An example of a rapidly migrating tidal inlet is pictured in Figure 34.

3) **A number of natural processes that remove**

sand from the immediate vicinity of the beach zone, such as: a) transport of sand offshore during storms beyond the depth from which the sand is normally returned to the beach during calm periods; b) wind transport of the sand landward out of the normal zone of the beach cycle; and c) sand washed over the barrier island into the backbarrier region during major storms.

There are also a number of man-induced changes that could produce a deficit of sand at a particular beach, including:

1) **Dams on rivers.** Sediment contribution of any river discharging on the coast is of importance to the sediment budget of that coast. If a dam is built on the river to create a reservoir for storage of water, the current velocities in the river are reduced to such an extent that practically all sandy sediment carried by the river settles in the reservoir. The water discharged from the dam contains very little sand; therefore, the sand supply formerly delivered to the coast by that river is cut off. This is the overwhelming cause of beach erosion on shores with narrow or absent coastal plains, such as Japan and California. Even in South Carolina, the construction of a dam on the Santee River in 1942, which created the Santee/Cooper lakes (famous for their largemouth and striped bass fishery), caused the beach to erode up to hundreds of yards at the river's mouth after the dam was built.

2) **Diversion of rivers.** Some engineering projects call for the diversion of rivers away from their original channels. In addition to the construction of the dam across the Santee River in 1942, approximately 90% of the river flow was diverted into the Cooper River, which flows into Charleston Harbor. This assured that even most of the fine-grained sediment carried by the river was diverted from the river's mouth. This also contributed to some extent to the erosion of the delta shoreline.

3) **Sand mining.** This is not a practice that is accepted on the beaches in South Carolina, but it is a common practice in many parts of the world. During a study of the shoreline of Oman, we observed how this practice contributed to the erosion of the sand beaches there. The extraction of sand from sand bars in the nearshore zone, such as ebb-tidal deltas, for beach nourishment projects is another practice that could be called sand mining. If such sand is left in its natural configuration, some of it might eventually end up on the shore and its removal could expose the beach to erosion in places. Even mining of sand from shoals located further offshore can possibly influence beach erosion. Dredging of these shoals may possibly have two side effects: a) alteration of depth contours, which changes wave refraction patterns and may cause a focusing of wave energy at different locations on the shore; and b) elimination of a natural breakwater, which acts to reduce wave energy arriving at the shoreline. We are presently working on a project sponsored by the U.S. Department of Interior, Minerals Management Service (MMS) to determine the depths at which such sand dredging of offshore shoals would cause the least harm.

4) **Tidal inlet stabilization.** As a general rule, the ultimate geomorphic form of tidal inlets is the result of an adjustment to the dynamic action of both tidal currents and waves. Also, the volume and pathway of littoral drift (volume of sand moving along the shore) contributes to the configuration of a tidal inlet. From the studies of O'Brien (1931), it is well known that the stability of an inlet is governed by the relative strength/volume of the littoral drift and the tidal prism volume (amount of water that flows in and out of the inlet during a single tidal cycle). Sand moving along the shore bypasses inlets by bar migration under the influence of waves or by tidal current transport of individual sand grains. On coasts with a large littoral drift, there is a tendency for the inlet to migrate downdrift due to the infilling of the mouth of the inlet on the updrift side.

Almost any man-made modification to

a tidal inlet, therefore, may have significant influence on the erosional/depositional patterns on adjacent beaches. For example, artificial deepening of an inlet to develop a harbor could change the bar-bypassing nature of the inlet, depriving the downdrift coast of its supply of sand and thus leading to erosion.

5) **Construction of shore perpendicular structures.**

Jetties, long parallel structures built at a river mouth or a tidal inlet in order to stabilize the channel, prevent its shoaling by littoral drift and protect its entrance from waves, are the most conspicuous of shore perpendicular structures. Jetties are also designed to direct or confine the flow to help the channel's self-scouring capacity. Usually, jetties extend through the entire nearshore zone to beyond the breaker zone in order to prevent the deposition of littoral drift in the main channel. Therefore, the jetties act as barriers to the longshore transport of sediments, causing a significant volume of the

sediment to accumulate on the updrift side of the jetties. At the same time, on the downdrift side, the sand transport processes continue to operate and cause the sand to move away from the jetties, resulting in the erosion of the shoreline on that side. In some instances, the updrift jetty may have a weir system whereby sand is trapped in a confined area so it can be either hauled offshore or passed across the jetties to the downdrift side of the inlet. The Charleston Harbor entrance jetties have a weir system, which will be discussed in more detail in Section II.

Groins are structures built perpendicular to the beach, in this case for the purpose of widening the beach by trapping a portion of the sand in the littoral drift. They are relatively narrow in width and may vary in length from about 30 to several hundred feet (see examples in Figure 57). Though their function differs from jetties, they are also barriers to longshore sand transport. If they trap a significant portion

FIGURE 57. Groins (arrows) at Edisto Beach, South Carolina (photograph by Tim Kana taken on 10 February 2006). This extensive groin field was installed between 1948 and 1975.

of the sand in transport, naturally the beach will erode on the downdrift side. Groins may be used in series (groin fields) to protect a large area, but the zone of erosion will shift down the coast through time. Once the sand builds out to near the end of the groin, some or all of the sand in transit will pass on downdrift. In some cases, engineers place nourished sand on the updrift side of the groins to prevent erosion on their downdrift sides. Poorly designed groins may deflect sand past their ends into deeper water, resulting in its loss to the longshore transport system. Kana et al. (2004) discussed the utility and management of groins on the South Carolina coast. They made eight planning guidelines and recommendations, including: 1) only use groins on the open coast where the erosion rate exceeds about 6 feet per year; and 2) construct the groins of a groin field in the downdrift to updrift direction, and nourish the "cells" between the groins "to capacity."

6) **Construction of seawalls and revetments.** Seawalls are vertical, hard structures very commonly built to protect man-made structures. Thousands of examples could be cited to demonstrate the effectiveness of such structures in saving property, at least in the short term. However, erosion of the beach itself is usually accelerated in front of these features because of wave reflection from the hard, vertical faces of the seawalls. According to Silvester (1977), when waves are obliquely reflected from such seawalls, energy is applied "doubly" to the sediment bed and "hence expedites the transmission of material downcoast." This process is illustrated in Figure 58. Over the past few decades, there has been a hue and cry by a number of concerned scientists against building seawalls along open ocean beaches, to the extent that some states now ban them (e.g., North Carolina). Revetments usually serve the same purpose as sea walls, but they are typically made of materials such as boulder-sized

chunks of rock, called riprap. These features do not reflect waves as severely as sea walls, but beaches seldom build out in front of them, for a variety of reasons, including wave reflection.

Sea-level Rise

When coastal geomorphologists like ourselves use the term sea-level rise, we normally insert the word **relative** as a modifier, because there are two possible factors that create major changes in sea level: 1) changes in the volume of water in the world ocean; and 2) the moving up and down of the earth's surface. Changes in the volume of water in the world ocean are mostly related to major changes in the world's climate. For example, during the peak of the last ice age, around 60,000 years ago, the surface of the world ocean was about 350 feet below its present level because so much of the earth's hydrosphere was taken up in the huge ice sheets that covered large areas of some of the continents.

A number of reasons can cause a land mass to rise or fall, including (to name but a few):

1) Tectonic activity such as mountain building. In areas of Alaska and Baja California, we have observed former beach lines elevated hundreds of feet up the sides of the mountains along the shoreline.

2) Elevation of the land as a result of the melting of the large ice sheets present during the Pleistocene Epoch, because the earth's surface was depressed under the heavy load of the ice. On Herman Melville Island in northern Canada, we observed numerous lines of beach deposits elevated many 10s of feet back of the shoreline as a result of a process called glacial rebound as the land rose slowly in response to the removal of the weight of the ice.

3) Sinking of land where large rivers have dumped huge sediment loads on the continental shelf. This kind of sinking, aided by compaction of

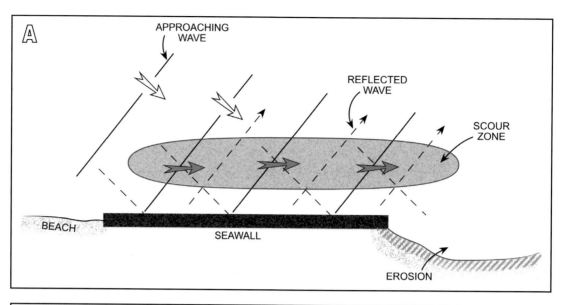

FIGURE 58. Seawalls and beach erosion. (A) Illustration of erosion in front of and on downdrift side of a seawall as a result of waves reflecting from the seawall (highly modified after Silvester, 1977, Figure 8). The approaching wave crests meet the reflecting wave crests at approximately right angles, generating a **flow parallel with the shore in the downdrift direction.** This current scours a channel a few yards seaward of the wall. The beach downdrift of the seawall will also erode due to loss of sand in the scour zone, which would normally accrete on the beach in the absence of such a strong current in the nearshore area. (B) Waves reflecting off a seawall at Hampton Beach, New Hampshire. Solid arrow indicates approaching wave and dashed arrow indicates reflected wave moving away from the seawall at a 90-degree angle to the approaching wave. Photograph by A.D. Hartwell taken circa 1969.

the sediments, has caused sea level to rise up to 4 feet per century in areas of Louisiana, where former lobes of the Mississippi River delta were abandoned as the river shifted positions. Once the delta lobe was abandoned, the river no longer supplied sediment to continue the delta-building process, and as a consequence, the abandoned lobe began to sink.

On the coast of South Carolina, which is located on the relatively stable trailing edge of the North American Plate, sea level is affected primarily by changes in the volume of water in the world ocean.

For the coast of South Carolina, available data suggest the following historical rates of sea-level rise (partly from a summary by Pilkey and Pilkey-Jarvis, 2006):

1) Between 15,000 and 6,000 years ago – Overall rate of **3 feet per century**, with possible "blips on the curve" of as much as 10 feet per century. Also, there were some fairly well documented stillstands during this rise, at which time river deltas and barrier island systems were deposited, the remnants of which are still present on the continental shelf in places (see Figure 8).

2) Between 6,000 and 4,500 years ago – Probably around **1.5 feet per century**.

3) Between 4,500 and the present – This is the time interval we refer to as the present "**stillstand**" of sea level. It is during this period of time that all of the modern deltas and barrier islands on the South Carolina coast have been formed. As shown by Figure 9, the curve showing sea-level changes during that time has not been a straight line, but has bumped up and down several times during this "stillstand" period, sometimes more than 3 feet.

Having been working on the South Carolina coast for the past 35 years, it is clear to us from simple observation that sea level is rising. How much? Probably around one foot per century at the present time.

The theory that a rising sea level is the major cause of beach erosion worldwide prevails in both scientific and popular literature (e.g., Concerned Coastal Geologists, 1981). A careful analysis of beach erosion in a number of settings casts considerable doubt on this assumption, particularly with regard to problems of concern in the near future. In fact, we think that, in most cases, man's impact on sediment supplies and unwise construction practices cause most beach erosion problems. However, the opinions of the experts differ markedly on this topic.

Our friend and coastal engineering *guru*, Cyril J. Galvin, challenged the famous report by Concerned Coastal Geologists (1981). Hayes was "concerned" about a lot of things in 1981, but, for some reason, he was not invited to the conference that generated this report. Anyway, in his challenge, Galvin stated "sea-level change has negligible effect on shore erosion, compared to fluctuations in longshore sand transport rate." On a site-specific basis, his assertion is demonstrably true. In South Carolina, for example, beaches that are eroding tens of yards per year are located within a few hundred yards of beaches accreting tens of yards per year. Such differences are usually related to sediment bypassing at major tidal inlets during which a sand deficit is created on one side of the inlet, as will be shown in the discussion of the individual sites in Section II. Obviously, both the eroding and depositing beaches in these instances were subject to the same conditions of sea-level change. On the other hand, changes in water level clearly impact shoreline erosion in the Great Lakes, which show 11-year cycles of major water-level changes (Hands, 1977).

Water-level changes not related to the aforementioned global patterns also affect beach erosion trends. On the South Carolina coast, with its considerable tidal range, the most severe erosion on the beach occurs at high tide. During spring tides, higher levels of the beach are exposed to wave action than during neap tides, so erosion is at a maximum during spring tides under similar wave regimes. Consequently, coastal storms usually cause maximum rates of beach erosion when they cross the coast during high spring tides.

Water-level changes also show seasonal variations. On the east coast of the United States, sea level is generally lowest in the spring and highest in the fall. This fact is well known by east coast developers who have built too close to the beach, as the high tides in the fall commonly cause them problems. According to Komar (1976), these annual sea-level changes can be attributed primarily to seasonal variations in climate and ocean water properties. However, much is yet to be learned about this process.

Why the issue of **global warming** and the potential of this temperature rise to cause sea level to continue to rise in the future is such a controversial political issue is a complete mystery to us, because the evidence is indisputable that: a) global warming is occurring, based on evidence such as melting glaciers (which we have witnessed first hand in Alaska) and actual temperature measurements; and b) the concentration of CO_2 in the atmosphere is steadily increasing, based on studies such as at the Mauna Loa Observatory in Hawaii over the past 40 years. The relationship of these two issues is treated in detail in Al Gore's book, *An Inconvenient Truth*.

Pilkey and Pilkey-Jarvis (2006) also discuss this topic in an informative way. They summarized the issue succinctly in the following statement:

"The evidence for the human connection to global warming is the correspondence of the massive production of excess carbon dioxide produced by burning fossil fuel in the last few decades and the simultaneous atmospheric increase in carbon dioxide. On a purely physical basis, the additional atmospheric CO_2 requires that some greenhouse warming must occur, but how much remains a question."

So what does the future look like with respect to sea level on the South Carolina coast? Pilkey and Pilkey-Jarvis (2006) have the following assessment that relates to that issue as well:

"The current most widely accepted prediction of sea-level rise is that its rate will be two to four times the present rate by the year 2100, and at that time the sea level will be two to three feet above its present state. At the same time, atmospheric temperatures will rise four to five degrees Fahrenheit. But these are numbers with a lot of leeway."

So much leeway, in fact, that the very morning we are writing this (29 January 2007), the local *The State* newspaper has an article on an upcoming meeting in Paris to finish the first of four major global warming reports by an international panel sponsored by the United Nations. As expected, this involves a lot of controversy. The early versions of the report in question predicted that by 2100, the sea level will rise between 5 and 23 inches, which is far lower than the 20-55 inches forecast in the same time frame in a scientific article published in the journal *Science* this month. That second number, as you can see, is close to what Pilkey and Pilkey-Jarvis had stated.

The problem, as with a lot of models used to predict change in nature, is the suppositions, or data input, used to create the model. In the earlier models, the data input related to rising sea level was the melting of glaciers, which is well documented, and the physical expansion of water as it warms. They did not take into account large-scale melting of the big ice fields in Greenland and Antarctica, which is pretty scary when you consider the following statement in the article:

"In 2002, Antarctica's 1,255 square-mile Larsen B ice shelf broke off and disappeared in just 35 days. And recent NASA data shows that Greenland is

losing 53 cubic miles of ice each year – twice the rate it was losing in 1996."

That type of major melting was addressed as one of the possible doomsday scenarios in Al Gore's book.

All of that being said, there is no doubt that an abrupt rise in sea level of two or three feet would wipe out an unimaginable amount of beach homes, hotels, golf courses, and other structures that are now built close to the beach in South Carolina. Question is, what can be done to prevent such a rise? The answer to that question is way beyond the scope of this book.

METHODS COMMONLY USED TO PREVENT BEACH EROSION

Engineers have attempted to curtail beach erosion for centuries. Their success ratio has been variable, depending upon the vagaries of the sediment supply, changes in water levels, and storm-wave conditions. For purposes of discussion, we divide the techniques used to prevent beach erosion into two classes, "hard" engineering solutions and "soft" engineering solutions.

In the past, engineers have usually dealt with beach erosion by building resistant, permanent features that reflect or dissipate incoming waves. Some of these "hard" solutions include:

1) **Seawalls, revetments, and bulkheads.** In places where fixed property, such as highways and large hotels, are threatened by erosion, these three types of features are commonly built. As noted earlier, seawalls, massive concrete structures designed to hold the line against storm-wave erosion (illustrated in Figure 58), are a double-edged sword, so to speak, because, while they tend to succeed in keeping the road or hotel in place, at least until a major hurricane hits, the beach sand itself is usually sacrificed, even under normal wave conditions. A bulkhead is made of pilings, composed of a wide range of materials, driven into the ground. They are usually built in areas of moderate waves. Revetments are constructed by armoring the slope or face of a dune or bluff with one or more layers of rock, concrete, or asphalt. The armor stones of revetments tend to dissipate waves and inhibit reflection better than vertical walls, thus sand removal by wave reflection is not as severe as it is for seawalls. These types of features are severely scrutinized these days by public officials in South Carolina. In fact, the Beachfront Management Act of 1990 prohibited new seawalls. However, a settlement reached on March 9, 2004 involving some land owners on Daufuskie Island appears to have allowed some to remain in place.

2) **Groins.** These features, which are discussed in some detail in the section on causes of beach erosion (illustrated in Figure 57), are commonly made of rubble stone, but they may consist of wood, sand bags, gabions (rocks or gravel in wire mesh), or other materials. They work best where waves approach the coast obliquely. Whereas they work well if installed properly in the right place, groins have not proved to be an effective solution to beach erosion in many localities, especially where the dominant wave crests approach parallel to the beach. Although they have been used in abundance, groins were banned originally by the South Carolina Beachfront Management Act of 1990. However, that Act was later amended to expressly allow groins, so don't be surprised if you see a few on South Carolina's sandy beaches. Also, of course, many were constructed before the Act was passed (see Figure 57 for an example of some).

3) **Offshore breakwaters.** These features, which are usually composed of riprap or heavy concrete blocks of miscellaneous shapes, are built offshore, detached from the beach (see Figure 59). They reduce wave energy on their

FIGURE 59. Offshore breakwaters (arrows) near Port Fourchon, Louisiana (photograph taken in the spring of 2005). The effectiveness of these breakwaters is hindered by the fact that the land is sinking at a relatively rapid rate, because of the abandonment of a lobe of the Mississippi delta several hundred years ago.

landward sides, which causes sand moving along shore in the longshore transport system to accumulate in their lee. Such an obstacle to the movement of the sand along shore will commonly cause erosion on their downdrift side (note the shoreline erosion beyond the end of the breakwaters in Figure 59). Thus, many engineers recommend placing nourished sand in the shelter of these structures so that sand can continue moving along the shore. Offshore breakwaters have been used successfully to curtail erosion in a number of areas, the ones on the shoreline of Israel come to mind. However, there are none on the South Carolina coast that we know of.

There are many workers in the area of coastal erosion, particularly coastal geologists, who prefer to use solutions to beach erosion that do not involve hard structures. Two of the lines of reasoning used to support this position are that hard structures, such as seawalls, accelerate sand loss, and once in place, hard structures are difficult to remove, making it virtually impossible to correct a mistake. Some examples of these types of "soft" solutions include:

1) **Setback lines.** The "softest" of the soft solutions is the construction a setback line behind the shoreline seaward of which building is prohibited. They work best and are easiest to implement in areas that have not been developed as yet. The criteria used to establish

such lines include historical analysis of shoreline trends, preservation of the line of foredunes, defining areas of flooding and storm wave uprush, and so forth. In 1974, Hayes and his associates mapped out setback lines for the developers of Kiawah Island, estimating a 50-year lifespan for the security of structures built behind them. The developers of that island have for the most part adhered to those setback lines and have not lost any significant coastal structures at the time of this writing, over 30 years later. More details will be given in the specific discussion of Kiawah Island in Section II.

2) **Sand bypassing.** As noted earlier, jetties constructed to stabilize navigational channels usually block the flow of sand along the shore, with severe erosion commonly developing downdrift of the jetties. This is one of the more common major causes of erosion problems around the world. This type of erosion problem may be at least partly solved by installing mechanical sand bypassing systems, such as land-based dredging plants, pumping systems, and so on, which move the sand from the updrift to the downdrift sides of the jetties. Such systems have been established in many areas, with several examples on the Florida coast. In South Carolina, jetties exist at Little River Inlet at the North Carolina/South Carolina border, Murrells Inlet, the entrances to Winyah Bay and Charleston Harbor, and the mouth of the Savannah River on the South Carolina/Georgia border. Only the Murrells Inlet jetties have a working sand bypassing program, as far as we know. There is no fixed-in-place bypassing plant at Murrells Inlet, but the U.S. Army Corps of Engineers has a schedule for periodic sand bypassing using a hydraulic dredge (T. Kana, pers. comm.).

3) **Relocating a tidal inlet.** Migrating tidal inlets erode the shore as they move down the coast. Relocating the inlet back up the coast in the direction from which it came would relieve the down shore area of erosion, at least until the inlet migrates back. To our knowledge, one of the major examples, and maybe the only one up to that time, of this was carried out by our group in March 1983 at Captain Sams Inlet on the South Carolina coast, when the inlet was moved to a new dredged channel approximately 4,000 feet up the coast (to the northeast). This stopped the erosion of the inlet into a development to the south, as well as provided a huge volume of sand to an eroding beach to the south when the abandoned ebb-tidal delta of the inlet was driven ashore by wave action. Compared to establishing hard structures or jetties, this was a relatively inexpensive process. The downside is that it has to be repeated about every 14 years or so, but it would take many repeats to reach the costs of other types of protection. This operation is described in more detail in the discussion of the Barrier Islands compartment.

4) **Beach nourishment.** Beach nourishment with sand is the most commonly used of the "soft" solutions. Numerous projects of this type have been carried out around the world, some of which involve moving millions of cubic yards of sand (e.g., at Miami Beach, Florida). Sources for such sand include dredging sand deposits on the continental shelf, dredging sand bodies associated with tidal inlets, moving sand by various mechanisms from adjacent beaches with an abundance of sand, and hauled or pumped in from land-based sources. The major objection to beach nourishment schemes is that they are not permanent. Watching millions of dollars worth of sand wash away within a few years does not appeal to either public servants or even casual observers. Therefore, such projects require a careful analysis of monetary costs and benefits balanced against the aesthetic, economic, and recreational value of maintaining a sand beach in place. This issue will be discussed in much

more detail in Section II.

The material presented up to this point is a far from complete list of the numerous methods that have been tried to stop beach erosion. But as you may have guessed by now, this issue goes beyond a simple engineering or geological problem, having blossomed into a socio-economic, political, and even cultural phenomenon. The fact that a large percentage of the world's population lives near the coast amplifies interest in the subject.

Over simplifying the story quite a bit, one could say that the scientists and engineers who are supposed to know something about beach erosion fall into two major camps. On the one hand, we have environmentally oriented marine scientists, particularly coastal geologists, who are more inclined to let nature take its course and not crowd the ever narrowing beaches with beach cottages, hotels, etc. They usually think most beach erosion is caused by rising sea level or bad engineering practices and are more or less convinced that the problem will get worse with an increase in global warming. One of the leading spokespersons of this faction is another one of our friends, Dr. Orrin H. Pilkey, emeritus Professor at Duke University, who has been known to use catch phrases such as "New-Jersey-fication of the coast" and "let the lighthouses fall into the sea." He has also authored and edited many books on living with the coast. He is not even much of a fan of some "soft" solutions, such as beach nourishment, heavily criticizing some of the models engineers use to predict the life of the nourished sand on the shore. Needless to say, Dr. Pilkey is not idolized by many coastal engineers. However, some of his arguments have been heeded by policy makers in different states of the U.S. (e.g., North Carolina's ban of seawalls on the open beach). If sea level does rise abruptly, many of Pilkey's arguments will be awfully difficult to refute.

And on the other hand, we have the coastal engineers, who are trying to perform a very difficult job. Our close friend, Bill Baird, a world renowned coastal engineer, has founded a company based in Canada (Baird and Associates) that has the following motto – "making the world a better place to be!" His company does that by designing ports in developing countries to fuel their economy, preventing coastal villages in Africa from falling into the sea, and so on.

Where do we stand on this issue? To put it simply, our hearts are with the notion that the beaches should be left free of so much development, but our heads love the science of dealing with the problem. There are some classic cases of mismanagement of the beaches in South Carolina, and we will point out a couple of them later in the detailed discussions of the different coastal compartments. We have owned and still own coastal property in the state, but never on the open beach. We were once offered a lot on one of the barrier islands as "payment in kind," but didn't take it because of concerns about erosion during hurricanes. However, when faced with a tricky scientific problem to solve, such as moving a tidal inlet or keeping an African village from falling into the sea, our heads win and we enthusiastically tackle the problem. In fact, we feel pretty good about some of the projects we have done around the world.

HURRICANE *HUGO* (1989)

No discussion of beach erosion on the South Carolina coast would be complete without a review of the impacts of hurricane *Hugo*, the eye of which crossed the South Carolina coast just northeast of the center of the city of Charleston near midnight on 22 September 1989 (see image in Figure 60 of the hurricane as it approached South Carolina at 2:44 pm. on 21 September). By conducting a number of field trips, research projects, and training courses along the coast every year since the storm, the authors have been able to observe its recovery. Fifteen years later, evidence of the effects of the storm is still visible in a few places, but in general, only the trained eye can recognize it. At that time,

Hurricane Hugo
2:44 p.m. EDT
September 21, 1989

FIGURE 60. Hurricane *Hugo* approaching the South Carolina coast on 21 September 1989. Image courtesy of National Oceanic and Atmospheric Administration.

however, this, the largest hurricane to strike the South Carolina coast in the 20th century, was most impressive.

Hugo originated as a classic Cape Verde hurricane, which was first detected as a tropical wave emerging from the coast of Africa on 9 September 1989 (as reported by NOAA sources). The storm gathered strength quickly, becoming a Category 5 hurricane (winds 156 mph and greater) on 15 September. The storm passed over St. Croix, USVI on its way toward Puerto Rico, passing over its eastern end on the 19th. After leaving Puerto Rico, the storm eventually headed directly for the South Carolina coast. Strengthening in the last twelve hours before landfall made *Hugo* a Category 4 hurricane (winds 131-155 mph) at the coast. After landfall, the storm gradually recurved

northeastward, becoming extratropical over southeastern Canada on 23 September (as reported by NOAA).

A ship moored in the Sampit River, 55 miles north of Charleston on the South Carolina coast, measured sustained winds of 120 mph. High winds associated with *Hugo* extended far inland, with Shaw Air Force Base, South Carolina (about 90 miles from the coast) reporting 67 mph sustained winds with gusts to 110 mph, and Charlotte, North Carolina reporting 69 mph sustained winds and gusts to 99 mph.

The storm surge from *Hugo* inundated the South Carolina coast from Charleston to Myrtle Beach, with maximum storm tides of greater than 20 feet observed in the Cape Romain-Bulls Bay area, approximately 30 miles north of where the eye crossed the coast (Figure 25). This was the largest storm surge experienced on the east coast of the United States in the 20th century.

Hugo was responsible for 21 deaths in the mainland United States, five more in Puerto Rico and the U.S. Virgin Islands, and 24 more elsewhere in the Caribbean. Damage estimates were $7 billion in the mainland United States and $1 billion in Puerto Rico and the U.S. Virgin Islands (also reported by NOAA).

Some of the details of the impacts of the storm on the different parts of the coast will be presented as the individual compartments of the coast are discussed in later chapters, but some generalizations are in order. Except for the developed island just southwest of Charleston, Folly Island, significant impacts of the hurricane were experienced only to the northeast of Charleston. Places like Kiawah Island, Beaufort, and Hilton Head were spared because of their location to the "left" of the storm's eye (see Figure 25). On the other hand, coastal structures were heavily impacted as far northeast as Myrtle Beach, because of the amplification of the storm surge on that side of the eye. Economic impacts to man-made structures would have been much greater had approximately 50% of the shoreline between Charleston and Myrtle Beach not

been undeveloped natural areas. The publication by Finkl and Pilkey (1991) is a compilation of a number of scientific articles presenting details on the impact of the storm, including one by Sexton and Hayes (1991).

Table 2 shows data presented by *The State* newspaper in Columbia with regard to the number of structures (mostly beach front residences) completely undermined and removed by wave erosion during *Hugo*. At that time, the Coastal Council of South Carolina had proposed a concept called the "primary dune," stating that one could not build a beach house on the seaward side of that feature. The "primary dune" was synonymous in most places with what we defined earlier as the **foredunes** behind the beach. This line of dunes was washed out for the entire length of the coast from just southwest of Charleston (on Folly Island) almost to the North Carolina border. With the disappearance of this dune line, as well as the sand underlying the dunes and the beachfront houses, a total of 159 of the pre-storm houses theoretically could not be rebuilt. This resulted in a long series of legal wrangling that resulted in South Carolina relaxing its restrictions on beachfront development. Surprise, surprise! Getting into the issues of the legality of coastal development and insurance

TABLE 2. Oceanfront structures destroyed during hurricane *Hugo* that had foundations (or eroded-out foundations) located seaward of the "primary dune" once the storm passed (modified after The State).

LOCATION	NUMBER	GEOLOGIC LOCATION
Folly Beach	32	Sediment starved prograding barrier island
Isle of Palms; Sullivans Island	6	Prograding barrier island
Pawleys Island	34	Landward-migrating (transgressive) barrier island
Garden City	66	Landward-migrating spit
Five other areas	21	Miscellaneous

regulations is well beyond the authors' goal for this book. We are not saying it isn't important, but just that the answers to these issues should be found elsewhere.

Table 2 illustrates the correlation of losses of beachfront dwellings to the **geomorphology** of their settings. For example, 32 of the structures that theoretically could not be rebuilt were located on Folly Island, the second island south of the entrance to Charleston Harbor. This island has had a chronic erosion problem for about the last 50 years, with many houses being lost in the pre-*Hugo* past during erosional episodes. This erosion was caused by the trapping and loss of sand from the normal longshore sediment budget due to the presence of the jetties that maintain the channel into the harbor. These jetties were completed in 1898, but it took many years for the erosion at Folly Island to become severe for reasons to be discussed in Section II. Before the jetties were built, Folly Beach was a normal prograding (seaward-building) barrier island; however, since the sand supply was cut off, the seaward side of the island has been eroding. In any event, erosion at Folly Island during hurricane *Hugo* was not surprising, considering its history.

Only six beachfront houses on the Isle of Palms and Sullivans Island, the islands just to the north of the Charleston jetties, were so classified (Table 2). Unlike Folly Island, these islands have continued to receive an adequate supply of sand from the northeast so that they have historically built out (prograded) seaward. Despite the relatively low number of house sites "now seaward of the primary dune," there was some erosion on the Isle of Palms, as is shown by the photographs in Figures 61 and 62. Because of the continued progradation of the shore of this island, the houses had enough of a set back from the high-tide line (and the primary dune when it recovered) to prohibit their foundations from being undermined.

Unlike the relatively light "permanent" loss of beachfront house sites on the Isle of Palms, 100 beachfront houses at Garden City and on Pawleys Island (in places) were eroded out and completely undermined to the extent that their original building sites were classified as "seaward of the primary dune." The reason for this disparity in comparison with the Isle of Palms is that beachfront houses at those two localities had been built on portions of barrier islands that historically were moving landward (transgressive barrier islands; see Figures 63 and 64). The photograph shown in Figure 63 of the south end of Pawleys Island taken two years and eight months before hurricane *Hugo* (1989) illustrates the crowded nature of the houses on a narrow spit on the southern end of the island. As the sketch map in Figure 64A shows, this low-elevation spit was completely washed over during hurricane *Hugo* and several of the beachfront cottages were transported over the tidal channel behind the spit and deposited on the salt marsh on the landward side of the channel. Also note that a new channel was cut through the spit. A channel was also cut through that same spit during hurricane *Hazel* in 1954, which had a storm surge of over 14 feet in the Myrtle Beach area. The sketch of a photograph taken soon after *Hugo* (Figure 64B) shows a house in the new channel. A vertical image of the south end of the island taken in 2006 shows that the spit area had been completely rebuilt with houses by then (Figure 65).

As with other major hurricanes in temperate or subtropical climates, hurricane *Hugo* had a major impact on the trees in the path of the storm. Most of the mature trees on the surface of Bull Island, which was subject to some of the strongest winds during the storm, were blown down. Seven counties in South Carolina in the path of the storm received major damage to the trees (see Figure 66). The South Carolina Forestry Commission listed the following damages to the forest resources of the state:

- A total of 36% of South Carolina's 12.2 million acres of forested land was damaged.
- The state lost enough timber to "house nearly the entire population of West Virginia."

FIGURE 61. Oblique southeasterly view of the central shoreline of the Isle of Palms taken a few days after hurricane *Hugo* (11 October 1989). Arrows point to locations where beachfront houses were completely removed from their foundations and transported landward.

FIGURE 62. Isle of Palms erosion during hurricane *Hugo* (1989). Arrows show path of migration of a house formerly on the front row of beach houses that was deposited in a neighbor's yard during the storm surge of the hurricane.

FIGURE 63. Southern end of Pawleys Island showing closely spaced houses on the washover terrace of the southern spit. Photograph by Tim Kana taken on 2 January 1987, two years and eight months before hurricane *Hugo* (1989).

- About 70-80% of the "saw timber" in the Francis Marion National Forest was lost, enough to make a 12-inch board reaching around the world 5 to 7 times.

In summary, the major physical impacts of the storm were:

- Extreme flooding because of a huge storm surge (greater than 20 feet in some places).
- "Expanded" tidal inlets with erosion on the adjacent beaches. Figure 67 shows a general model of this type of "expansion."
- Abrupt landward motion of barrier islands –

maximum beach property damages observed in those areas.

- Elimination of the foredunes for 10's of miles along the coast.
- Massive wind destruction of vegetation, especially of pine trees.
- Formation of new relatively small tidal inlets (e.g., across the necks of spits).
- Most damages occurring to the "right" of the eye of the hurricane.
- Shoals outside the entrance of the major tidal inlets (*ebb-tidal deltas*) planed off flat.
- Effects on the continental shelf, including scour channels and obvious offshore sediment transport (Gayes, 1991).

The South Carolina Wildlife Commission reported the following damages to the wildlife resources of the state:

- Sea turtles and shorebirds were hit hard.
- Eighteen of 19 bald eagle nests in the path of the storm (40% of all the nests in the state) were blown down.
- Many of the animals associated with old-growth forests were affected. For example, the red-cockaded woodpeckers lost 75% of their nesting trees (23% of all nesting colonies in the world).
- The 20-foot surge completely washed over Bull Island, a key component of the Cape Romain National Wildlife Refuge. Amazingly, the five red wolves that had been stocked on the island survived.
- There were massive fish kills in several coastal streams because of low dissolved oxygen.

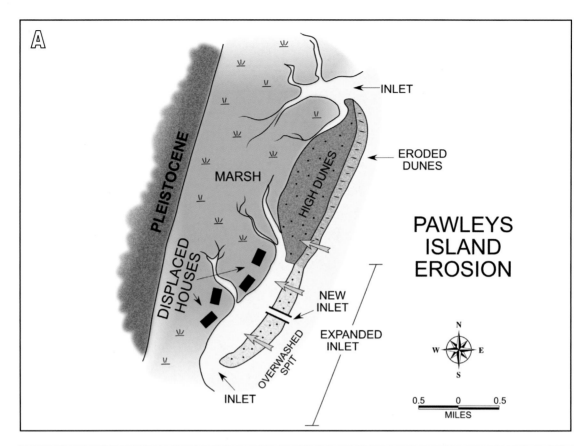

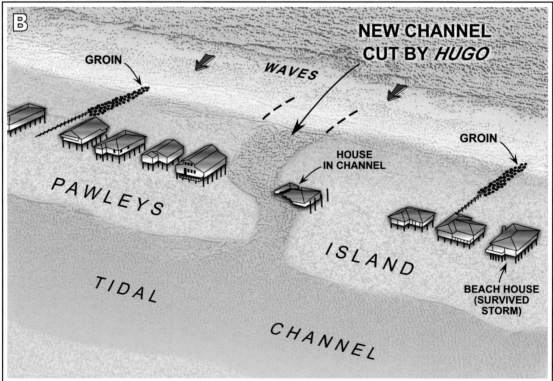

FIGURE 64. Erosion at Pawleys Island during hurricane *Hugo* (1989). (A) Schematic showing location of displaced houses. (B) House in new channel cut by the storm. Based on photograph by Bill Jordan.

FIGURE 65. Image of southern end of Pawleys Island acquired 2006, courtesy of South Carolina Department of Natural Resources. Many of the houses on the spit were built back after hurricane *Hugo* (1989).

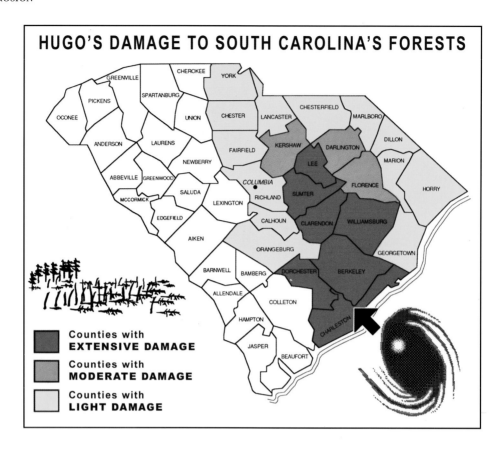

FIGURE 66. Damage to the trees in the forests of South Carolina as a result of hurricane *Hugo* (1989). Courtesy of the South Carolina Forestry Commission.

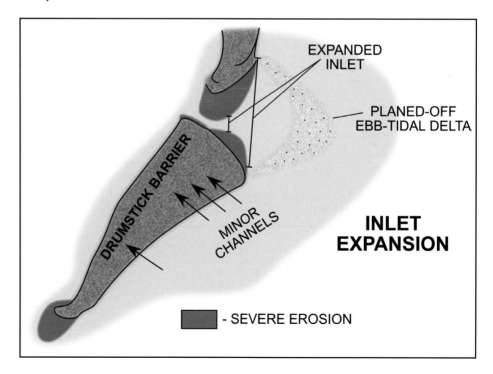

FIGURE 67. Generalized model for the expansion of tidal inlets observed on the South Carolina coast as a result of hurricane *Hugo* (1989).

7 ESTUARINE AND BACKBARRIER HABITATS

GENERAL MORPHOLOGICAL MODEL

Figure 68 shows a very general morphological model for the environments in the backbarrier and sheltered estuarine regions in South Carolina. There are numerous variations and exceptions to this general picture, particularly for those areas located in the zone of the turbidity maximum. In many estuaries, however, these areas can be subdivided into three distinct environments: the *salt marsh* (from mean sea level to the highest spring tide level); the *tidal flat* (spring low tide to mean sea level); and *tidal channel* (low spring tide level to the bottom of the channel). The marsh is commonly divided into low marsh and high marsh, and we could similarly divide the tidal flat into low and high segments. Generally speaking, the lower part of the tidal flat is subject to at least a modicum of turbulent water motion generated mostly by tidal currents and occasionally by waves under storm conditions. That part of the flat typically contains some sand exhibiting evidence of sediment transport (e.g., ripples). The high flat contains finer-grained sediment that is riddled with the burrows of multitudes of infaunal organisms, particularly polychaetes. Also, mud snails occur in abundance over the surface of the flats, leaving conspicuous feeding trails. The uppermost part of the flat (still below neap high tide) typically contains numerous

oyster mounds. As the mud accumulates around the oyster mounds, salt marshes may grow over the mud.

MARSHES

Introduction

The estuaries of the South Carolina coast are host to extensive developments of salt, brackish, and freshwater marsh systems, which are well documented as being the primary food sources for the coastal and nearshore ecosystems of the region. Marshes, like bottomland swamps, are flooded with water on a regular basis. The distinction between the two wetland types is that **herbaceous plants** populate marshes and **woody plants** populate swamps.

Marshes in estuarine systems occur in areas that were originally tidal flats where sediment accumulated in the early stages of evolution of the system. As the flat was built up to or slightly above mean sea level, marsh grasses began to take root. Once the grasses grew on the flat, the sedimentation process was accelerated because of the baffling effect of the plants. These present marshes are, in effect, intertidal flats vegetated with **halophytes** (plants that are adapted to grow in salty soil; Basan and Frey, 1977). Marshes can expand very rapidly, up to

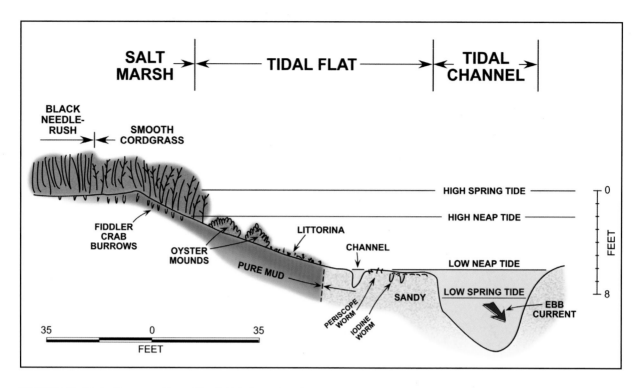

FIGURE 68. Typical topographic profile of tidal channel and associated habitats of the backbarrier and lower estuary regions in South Carolina.

a yard or so a year, if the slope is flat and sediments are abundant.

For purposes of description, we have divided the typical estuary in South Carolina into upper, middle, and lower (entrance) zones (Figure 40). As noted earlier, each of these zones is characterized by distinct hydrodynamic and sediment characteristics. **Freshwater marshes** are most common in the upper estuary, where there are tides but very low salinities; **brackish marshes** occur most commonly in the middle estuary, where salinities generally average less than 15 parts per thousand (ppt); and **salt marshes** occur in the lower estuary, where salinities range from 15 ppt to the low 30s.

These lateral salinity changes up and down the estuary have a striking impact on the plant communities. Giant cordgrass (*Spartina cynosuroides*) is conspicuous along the banks of the channels in both the upper and middle estuaries. Black needlerush (*Juncus reomerianus*) is by far the most common plant in the middle estuary, covering many thousands of acres in some places, and smooth

cordgrass (*Spartina alterniflora*) typically dominates the lower estuary. See Figure 69A for a delineation of the distribution of the more conspicuous plants throughout the estuaries of the Georgia Bight and Figure 69B for sketches of some of the dominant types. This discussion applies only to those estuaries that have a significant amount of fresh water coming into them (e.g., ACE Basin estuaries associated with the three main rivers entering them; Charleston Harbor's extended estuaries). Exceptions occur where freshwater input is only minimum or at the mouth of the major rivers.

Freshwater Marshes

In the upper reaches of some estuaries, as well as along the lower reaches of some of the major rivers, grassy wetlands composed of freshwater vegetation form a border along the major channels. The substrate in these wetlands is seldom exposed, because daily water-level changes are small to nonexistent, but, of course, they can be completely

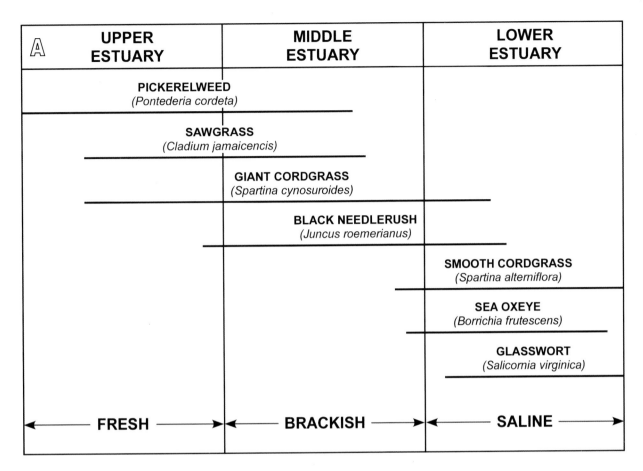

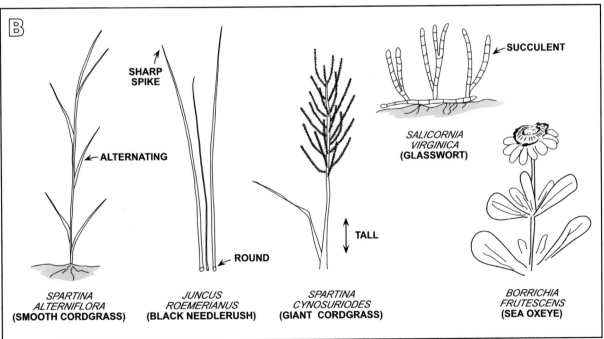

FIGURE 69. Marsh plants. (A) Occurrence of the most conspicuous plants in the marshes of the upper (fresh), middle (brackish), and lower (saline) parts of the estuaries of South Carolina. Modified after Stalter (1974). (B) Sketches of some of the more common marsh plants in South Carolina.

covered during floods along the river. In the upper estuaries and at the heads of the Santee/Pee Dee delta, for example, broad wetland areas populated predominantly by sawgrass (*Cladium jamaicencis*) are present.

Brackish Marshes

The brackish marshes of the larger estuaries of South Carolina, which, by definition, have significant freshwater influx, are dominated by black needlerush (*Juncus roemerianus*). These marshes have filled in broad expanses of the middle estuary regions, crowding in steeply against the banks of all the rivers and tidal channels. Tidal flats, which were abundant during the early phases of the evolution of the estuary, being in the zone of the turbidity maximum, are notably absent, having been covered by the grasses in most instances. Sediments in these marshes are, not surprisingly, mostly composed of silt and clay, which are highly mixed (bioturbated) by burrowing crabs, such as the fiddler crab (*Uca pugnax*) and by the roots of the marsh grasses.

Salt Marshes

The lower reaches of the estuaries and landward margins of barrier islands of the South Carolina coast are bordered by salt marshes dominated by smooth cordgrass (*Spartina alterniflora*). According to the South Carolina Sea Grant office, there are 504,450 acres of salt marsh in South Carolina (20 percent of the East Coast's total). Figure 70 shows a typical cross-section of a Georgia Bight salt marsh, based on the salt marshes in Georgia. Most experts agree that plant distribution in the marshes is controlled by depth and duration of flooding (Barry, 1980); therefore, it is convenient to divide these marshes into regularly flooded **low marsh**, the zone between mid-tide and neap high tide, and an irregularly flooded **high marsh**, which occurs roughly between neap and spring high tides.

Spartina alterniflora is the only plant that normally occurs in the low marsh zone, which is flooded 2-14 hours/day and has soil salt concentration of 0.5-3.2% (Barry, 1980). The higher portions of these plants can only be completely flooded about 10% of the time (else they die; T. Kana, pers. comm.), but the base of the lowest plants,

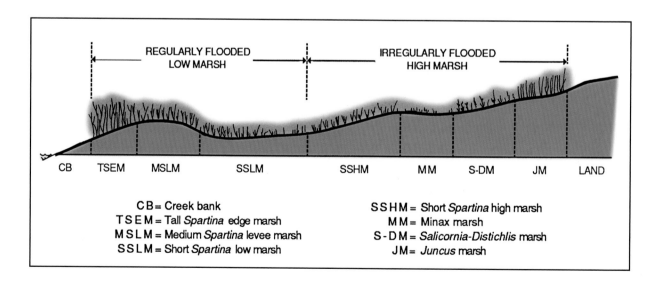

FIGURE 70. Typical profile of the salt marshes of the Georgia Bight (modified after Teal, 1958). The "Minax marsh" zone is named after the dominant species of fiddler crab found there, *Uca minax*. The dominant plant is extremely short *Spartina alterniflora*.

which are located near mean sea level, are flooded much more frequently. The *Spartina* is usually quite tall (greater than 6 feet at full growth) in the lower half of the profile (e.g., on creek banks and levees), but becomes dwarfed (0.3-1.7 feet) in the higher areas (e.g., between drainage creeks; behind levees; see Figure 70). The reason for these differences in height is still a matter of conjecture, but is thought to be related to the degree of flushing of the marsh soils. The vegetation in the lower marsh and creek banks is tall because the soils are better flushed by frequent tidal inundations.

In the high marsh, zones of glasswort (*Salicornia virginica*), sea oxeye (*Borrichia frutescens*), and salt grass (*Distichlis spicata*) grow in soils with salt concentration of 0.3-3.0% that are usually flooded daily but high enough in elevation to where the soils are flushed by rain water. The photograph in Figure 71 shows some of the marsh zonation in the backbarrier region of Kiawah Island. The upper high marsh, which is flooded mostly during spring tides, may be populated by black needlerush (*Juncus roemerianus*) in areas of lower salinity, and plants such as saltmeadow cordgrass (*Spartina patens*), marsh elder (*Iva frutescens*), or sea myrtle (*Baccharis* sp.) in more saline areas.

Most marsh sediments are low in oxygen, or reduced (usually dark brown to black in color; Wiedemann, 1972), except in topographically higher areas, such as levees and the high marsh, where water circulation at low tide via burrows and rootlets promotes oxidation (reddish color in some areas). The typically intensive reducing conditions in the sediments help account for the low species

FIGURE 71. Ground view of salt marsh surface in the backbarrier region of Kiawah Island. Note zonation of plants, with *Borrichia* occupying the high ground on the right and *Spartina alterniflora* occupying lower ground in the distance. *Salicornia* occupies the middle zone. Bass Creek, the major tidal creek landward of the north end of Kiawah Island, can be seen to the far right. Photograph taken in October 1978.

diversity in the salt marsh.

Mounds ("reefs") of the eastern oyster (*Crassostrea virginica*) play a prominent role in initiating salt-marsh growth in the lower portions of many of the estuaries. On tidal flats exposed to onshore flood-tidal currents, the oyster mounds grow in a row, oriented perpendicular to the current, presumably to take maximum advantage of availability of food particles to be filtered from the flowing water. The mounds are positioned so that their tops are just below neap high tide; therefore, at water levels higher than neap, they act as a baffle to the flood current, producing a current shadow on the landward side. Eventually, a mud flat builds up behind the linear mounds, over which the marsh can grow, and the oysters are eventually buried under the marsh. Oyster mounds also tend to grow across the entrances to tidal creeks draining the salt marsh (Ward and Domeracki, 1978). The dammed creek channel fills up with mud rapidly and the marsh grows over it, totally obscuring the old channels. Exposures of oyster mounds buried in this fashion and later exhumed by migrating tidal creeks are commonplace in the creek banks throughout the region.

TIDAL FLATS

Definition

Tidal flats are nearly flat intertidal surfaces that are relatively sheltered from direct wave attack. In most places, the sediment grain size decreases landward, away from main tidal channels (Figure 68), because the sediments are carried to their final resting place by tidal currents that decrease in velocity away from the channels as the water flows up onto the flats. The depositional and erosional rates on tidal flats are much slower than on beaches on the open-ocean front, where the sediment is moved readily by wave-generated currents.

Exposed Sandy Tidal Flats

In the entrances to some of the larger estuaries in South Carolina (e.g., entrance to St. Helena Sound), broad intertidal flats up to several hundred yards long and wide occur (Figure 72). These flats are usually composed of sand, which indicates that tidal currents and waves are strong enough to mobilize the sediments from time to time, removing the muddy components. Sedimentation rates are considerably slower than on the more exposed beaches. On the lower portions of some of these flats, waves may have formed low ridges that migrate very slowly across the flat and ripples and even larger bedforms may be present in the vicinity of tidal channels that cut across the flat, such as the one pictured in Figure 72. Despite this kind of sediment motion lower down, most of the sand and muddy sand on the upper half of these flats is relatively stable; therefore, it usually contains a huge population of infaunal organisms (those that burrow in the sediment). One of the more conspicuous burrowers is the Atlantic horseshoe crab (*Limulus polyphemus*; Figure 73). Others include numerous polychaeta, sparse clams, fiddler crabs (*Uca pugilator*), and the U-shaped burrows of the iodine worm, *Balanoglossus*.

A huge population of **horseshoe crabs** congregates in the spring on the large sand flat at the entrance to St. Helena Sound shown in Figure 72 for their annual reproduction ritual. This is no doubt true for other sand flats in the state, but we have observed them numerous times in the spring on this particular sandy tidal flat, because we always visited this flat at low spring tide during our field seminars. These fascinating creatures, which are pictured in Figure 73, are sometimes referred to as "living fossils," because fossil evidence indicates that they are over 300 million years old, preceding the arrival of the dinosaurs. They feed mostly at night, burrowing straight down into the sediment in search for worms and mollusks, including razor clams. They grow by molting and emerge 25% larger each molt.

Spawning occurs each spring during high tides of the new and full moon. At that time, males cluster along the water's edge as females arrive during the rising tide. Using glovelike claws, the male, which is smaller than the female, hangs on to the female's shell while she pulls him up the tidal flat toward

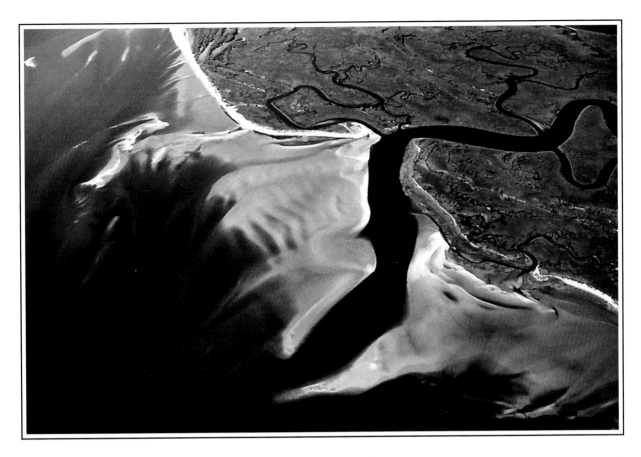

FIGURE 72. Intertidal sand flat exposed at low spring tide. This sand flat, located at the entrance to St. Helena Sound, is several hundred yards wide. Linear streaks in lower portion of image are plumes of suspended sediment being transported out of the sound by ebb currents. The multiple parallel sand bars in the middle of the image are evidence that this tidal flat is subject to some wave action. The mouth of the Ashepoo River is a short distance beyond the upper left corner of the image. Infrared photograph taken in December 1979.

the high-tide line (see Figure 73B). Once near the high-tide line, the female digs a hole every few feet to deposit eggs (up to 20,000). As the male is pulled over the nest, he fertilizes the eggs.

Not only are these creatures fascinating in their life style, but they also serve a number of important functions, including:

- Their eggs are an important food source for shorebirds, fish, and sea turtles. During the spring migration, several species of shorebirds, numbering in the thousands, time their arrival on the tidal flats of Delaware Bay to match the time when the horseshoe crab eggs are available. The Delaware Bay example is famous in the birding literature.

- The blue blood of horseshoe crabs contains *Limulus Ameboycte Lysate*, which detects harmful bacteria in drug products and medical devices.
- They play a major role in other aspects of medical research (cancer and vision studies).
- Indians of Roanoke Island, North Carolina tipped their fishing spears with the tails of horseshoe crabs.

Sheltered Mud Flats

Tidal mud flats occur under the following conditions on the South Carolina coast:

- **Sheltered reaches of estuaries that have rela-**

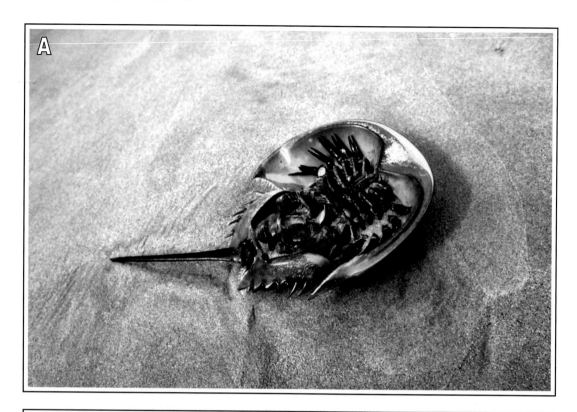

FIGURE 73. Horseshoe crabs (*Limulus polyphemus*) on the sand flat illustrated in Figure 72 during the spring mating season (April/May). In both cases, these horseshoe crabs were stranded on the flat at low tide. (A) Overturned on its back to show the underparts. This one was carried down to the waters edge and released. (B) Male mounted on top of a larger female, a typical mating arrangement. The female burrowed into the sand a small distance after the tide went out. Photographs taken circa 2003.

tively slow sedimentation rates. Two of the largest expanses of mud flats in South Carolina are at the head of Bulls Bay and throughout North Edisto Estuary. Figure 74 is a low-tide photograph of one of the mud flats in the North Edisto Estuary. Both of these areas are located far from freshwater river mouths; thus, the mud flats have not yet built up to levels that salt marshes can grow on them. Areas such as St. Helena Sound and most of the Georgia estuaries have adequate suspended mud available in the tidal waters to have built up the flats enough to allow marshes to cover them, so mud flats are somewhat sparse in those areas at the present time.

- **At the extreme watershed position in the lee of barrier islands.** Several of the barrier islands in South Carolina have barren tidal flats behind them at the position where the tidal flows from

the inlets on either side of the island meet (see photo and sketch in Figure 75). No detailed work has been carried out by us to determine the cause of this phenomenon; however, we speculate that more favorable growing conditions for *Spartina alterniflora* near the inlets result in taller plants and more efficient trapping of incoming suspended sediments in that area. Hence, the watershed areas (where the "tides meet") fill in last. These "bare spots" are rare on the Georgia coast, where suspended sediments are more abundant than on the South Carolina coast. Thus, we assume that the process of backbarrier filling by mud has taken less time on the Georgia coast (i.e., less than the 4,500 years since sea level stopped rising abruptly) than on the parts of the South Carolina coast that do not have significant freshwater river systems.

FIGURE 74. Extensive mud flat in the North Edisto Estuary at low tide. View looks northwest. Arrow points to line of oyster mounds. This flat is several hundred yards across. Infrared photograph taken on 2 May 1981.

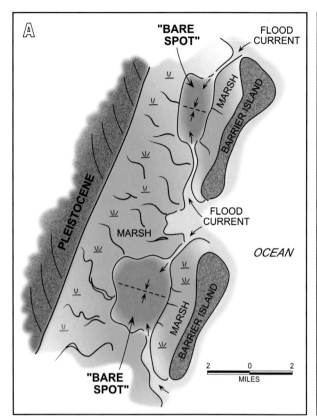

FIGURE 75. Location of "bare spots." (A) General model for the formation and location of "bare spots." (B) Location of "bare spot" in the lee of Bull Island (arrow). Image taken in 2006, courtesy of SCDNR.

- **Filling of abandoned channels**. Because of the fairly intense meandering mode of many of the tidal creeks and rivers on the South Carolina coast, they commonly abandon meander loops that become filled in with mud flats, which eventually are grown over by marshes. The process of channel abandonment, which usually takes a few decades in the smaller creeks, is constantly being repeated throughout the estuaries and backbarrier regions all along the coast (Figure 76A), which allows for the accumulation of mud flats such as the one shown in Figure 76B.
- **Channel margins, channel confluences, and tidal point bars**. Smaller tidal mud flats occur along channel margins, where a decrease in slope occurs, or at channel confluences, as a result of deposition down current away from

the confluence, which is typically a zone of bottom scour. Shallow portions of tidal point bars also contain minor mud flats in some localities.

The problem of why large quantities of mud accumulate on the upper reaches of tidal flats is an interesting one. Several Dutch workers have commented on this issue (Van Stratten, 1950; Postma, 1967), concluding that the mud is deposited during the last stages of the flood flow, during slack water, or at ebb just before water is drawn away, and it requires a stronger current to pick the mud back up than the one that deposited it. The finest particles in the mud are composed of clay minerals, which have a thin, sheet-like shape (like a thin coin). This shape slows down the process of settling of the particles from suspension,

FIGURE 76. Tidal creek in the Kiawah Island backbarrier region. (A) Meandering channel. Note how the former meander bend had truncated an older beach ridge, probably around 3,000 years old. Since the meander shown in this photograph was cut off, approximately 20 years before the photograph was taken, the abandoned oxbow had filled in with fine-grained sediment (arrow). Infrared photograph by L.G. Ward taken in the spring of 1978. (B) Ground view of the abandoned oxbow cutoff shown in photograph A. Note the abundance of fine-grained mud sediments that had filled in the oxbow. Marsh grass (*Spartina alterniflora*) fringes the channel. Photograph by A.W. Duc taken in May 1979.

which allows them to be transported landward by the waning flood currents beyond the point where more spherical particles would settle out. Also, once the thin, disc-shaped particles have settled on the bottom, their thin edges provide little resistance to the outflowing water; therefore, they are not easily picked back up into suspension. This processes of permanent deposition is aided further by the fact that floccules of organic matter and suspended fine-grained sediment (see Figure 40B), as well as slime-secreting diatoms, create other particles (in addition to individual silt and clay particles) that readily settle out in the quiet water. Once this particular mud is deposited, it tends to remain in place as well because: 1) diatoms move up through the mud and deposit slime; 2) the mud dries out, and 3) burrowing organisms tend to stabilize the mud. In some cases, mud is trapped between plants on the flat.

Another factor of prime importance to mud deposition is the super abundance of **suspension-feeding organisms**, such as oysters and clams. During the filtering process, these organisms compress the finely divided clay particles and bind them together in their intestines as fecal pellets,

which are excreted onto the flats. Another part of the suspended matter is coagulated in their gills and pushed back into the water as pseudo-faeces. These faeces and pseudo-faeces are easily deposited, even in comparatively turbulent water. This process is clearly evident around the oyster mounds throughout the estuaries and backbarrier regions in South Carolina.

Dominant Biogenic Features

The tracks, trails, and types of burrow structures of the animals that live on tidal flats and in their sediments have fascinated geologists for decades, because such evidence is useful for interpreting the depositional environments in ancient sediments (commonly used by geologists interpreting rock layers in their search for economic deposits, such as oil and gas). Knowledge on this topic has been greatly advanced by studies by our neighbors to the south (Georgia coast), specifically by associates at the University of Georgia (e.g., Fry and Howard, 1969; Howard and Dorjes, 1972). Their work provides a useful guide for the features you may see as you walk the intertidal regions of the South

Some of the more conspicuous biogenic features (commonly referred to as **biogenic structures**) to be seen in the sediments of the backbarrier and estuarine areas of the South Carolina coast include:

Environment	Biogenic Structures
1) High mud flats and sheltered portions of tidal point bars	Miscellaneoiu Polychaeta tubes (e.g., *Onuphis cremita)*; trails of snails (e.g., *Littorina irrorata)*
2) Higher portions of point bars and sandy tidal flats	Burrows of fiddler crabs (e.g., *Uca pugilator)*
3) Lower energy, water-saturated portions of sand flats and point bars	U-shaped burrows of the iodine worm (*Balanoglossus)* and vertical tubes of the periscope worm (*Diopatra cuprea)*
4) Exposed sand flats	Burrow pits of the horseshoe crab (*Limulus polyphemys)* and ghost shrimp (*Callianassa major)* burrow networks

Carolina coast (WITH SHOVEL IN HAND!!).

Figure 77 shows sketches of some of the more common burrow types. Except in some lower reaches of some sand flats and tidal point bars, where highly mobile, large-scale bedforms occur, the tidal flat sediments of the backbarrier and estuarine regions of South Carolina are highly burrowed throughout.

TIDAL CHANNELS

Before delving too far into this subject, a brief review of channel systems in general is in order. On a global scale, there are four basic styles of channel morphology that occur in natural river and tidal systems, which are illustrated in Figures 78A and 78B. **Straight channels** are rare in general, so be suspicious if you see one in the coastal region of South Carolina, because it has most likely been dug for navigation or other purposes, such as mosquito control or connections to old rice ponds. As a rule, **braided channels** occur under conditions of: 1) relatively steep channel slope; 2) high bedload content (sedimentary material on channel bottom; e.g., gravel and coarse sand, that moves along the bottom by rolling and bouncing); and 3) flashy discharge (high flow during flooding periods and low flow during the rest of the time). None of the conditions that result in braided streams occur in the extreme within the confines of the present lower coastal plain of South Carolina, so don't expect to see any braided streams as you sojourn there. However, had you been able to visit this area 12,000 years ago, all of the major stream systems would most likely have been braided, because of the steeper slopes (sea level about 350 feet below present) and availability of coarse-grained sediment because of those steeper slopes. **Meandering channels** tend to have: 1) flatter slopes than braided channels; 2) a high ratio of suspended to bedload sediments; and 3) a more steady discharge. Figure 79 shows a photograph of a meandering river on the coastal plain of South Carolina. Meandering channels are by far the most common channel type in the estuaries and backbarrier regions of South Carolina. **Anastomosing channels** typically occur

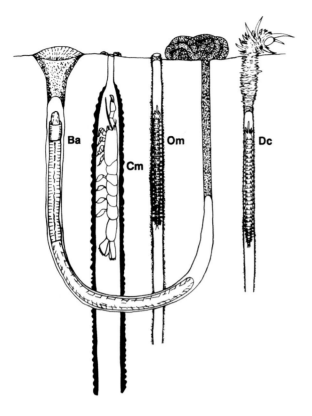

FIGURE 77. Burrowing characteristics of four of the more common organisms on the tidal flats of the South Carolina coast.

Ba – The iodine worm (*Balanoglossus*), which has an open u-shaped burrow. Note the excreted sand pile, a conspicuous feature on the sandy tidal flats where this organism lives.

Dc – The periscope worm (*Diapatra cuprea*), which lives in a vertical chitinous tube with a periscope-like entrance.

Om – The soda straw worm (*Onuphis microcephala*), which lives in a vertical mucoid tube covered with sand grains.

Cm – The ghost shrimp (*Callianassa major*), which lives in a complex vertical to horizontal catacomb-like network of burrows lined with mud-rich fecal pellets. This burrow is analogous to *Ophiomorpha* of the Cretaceous deposits of the Rocky mountain region.

(Modified after Howard and Dorjes, 1972.)

on extremely flat slopes in waters with a very high ratio of suspended to bedload sediments. A few anastomosing channels do occur on the South Carolina coast, but their origin is masked and made complex by systems with two opposing current directions due to the tides.

Considering only the tidal channels in the major estuarine systems, they are all located in the valleys carved during the last lowstand of sea level (called **lowstand valleys**). As noted earlier, sea level started to rise about 10-16,000 years ago and reached near its present level around 4,500 years ago (you have read these numbers before, but we have learned the hard way that these are numbers that bear repeating). With the relative stabilization of sea level, the old valley was filled up with sediments to a near-level plain that is now mostly covered with marsh grasses or tidal flats in some situations (through which the channels migrate).

The flat topography of the filled valley surface and associated gentle slope of the river channel, as well as the presence of the suspended muds of the zone of the turbidity maximum, dictate that tidal channels in the most landward reaches of all the estuaries and the two deltas on the coast are **meandering**. A perfect example of this, the most landward reaches of the May River estuary, is illustrated in the color infrared vertical photograph in Figure 80. As it turns out, meandering is a complex process, but, as noted above, it is clearly associated with flat stream slopes. Because of the combined effect of constriction of flow by the channel sides and bottom and the Coriolis effect, the pattern of flow of water in most channels is in the form of a helix. This results in alternating erosion and deposition of the channel sides in a downstream direction, which ultimately produces a meandering stream. Guess who first made this observation? Albert Einstein!

A sketch map of the morphology of a typical meandering tidal channel in the upper reaches of the estuaries on the South Carolina coast is given in Figure 81A. As the sketch shows, the outside bend of a meander usually has an associated **cut bank,** where the current undermines the bank and causes it to erode. Typically, the deepest part of the channel is located near the cut bank, where the maximum amount of erosion is taking place. An

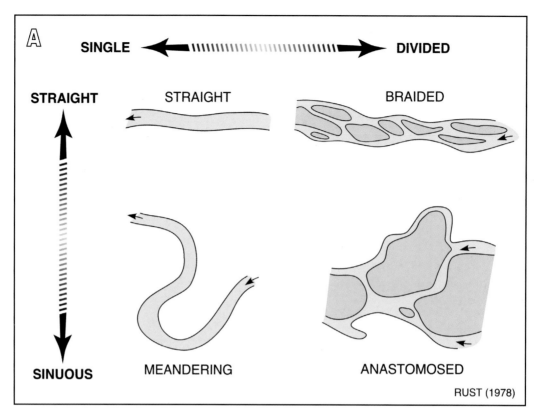

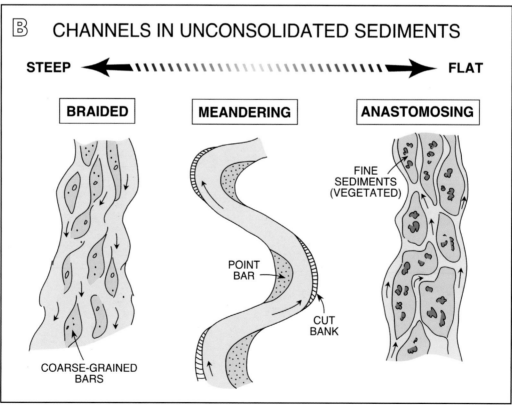

FIGURE 78. Channel types. (A) Classified as to whether they are straight or sinuous and whether they are single channels or a complex divided channel (after Rust, 1978). (B) Contrasting steep and flat channels.

FIGURE 79. Confluence of Congaree and Wateree Rivers (central South Carolina). Note complex meander belt of the Wateree River in the foreground. Arrow points to the bridge where US 601 crosses the Congaree River. Infrared photograph taken in September 1979.

arcuate-shaped bar, commonly composed of sand, called a **point bar**, occurs on the opposite side of the channel from the cut bank. The main force in creating this channel configuration in the upper estuary is the ebb current, which is stronger than flood currents in that part of the estuary. The reason for this is related to the physics of the tidal wave's progression, which is less efficient (i.e., takes longer) on the flood than the ebb. Therefore, even with equal volumes of water being exchanged on the flood and the ebb, mean ebb discharges need to be higher in order to accommodate the relatively shorter ebb current duration. Also, in most estuaries, and in both of the deltas, the freshwater runoff adds to the velocity of the ebb current. Another factor is the timing of peak flood and ebb currents. Peak flood currents occur late in the flood cycle (between mid

and high water) when the water is spread out over a wide area. Peak ebb currents occur near low tide, when the water is more confined to the channels themselves.

Further down, or more seaward, in estuaries, the flood current plays a more important role in channel configuration, resulting in the morphology illustrated in Figure 81B. In the middle and lower reaches of the estuaries, a bar tends to form near the middle of the main channel and a shallow channel on the side of the bar away from the steeper cut bank at the bend of the channel is dominated by flood currents from mid to high tide. The main channel near the cut bank tends to be host to somewhat stronger ebb currents than flood currents (for same reason as in the upper estuary). As a general rule, the middle estuary channels do not migrate

117

FIGURE 80. Vertical infrared image of upper May River, South Carolina acquired in 1994. Arrow points to location where the meandering channel has eroded into the adjacent Pleistocene upland. Image courtesy of SCDNR.

as rapidly as the more highly meandering channels further up the estuary.

Large point bars occur in the middle to lower estuary in some of the larger tidal rivers. A photograph of one of these **tidal river point bars** is given in Figure 82.

Near the estuary entrances, the channels are much wider and the ebb and flood currents are nearly equal. The main channels at that location are straighter than elsewhere and linear sand shoals tend to occur near the middle of some of the channels.

In the middle and upper reaches of the estuaries, in places, the meandering channels migrate against the valley walls of the original lowstand valley, which at this time is usually a forested upland. As development pressures from the construction of vacation and retirement homes on the coast have mounted, houses have been built on the sides of the old valleys at elevations normally considered

118

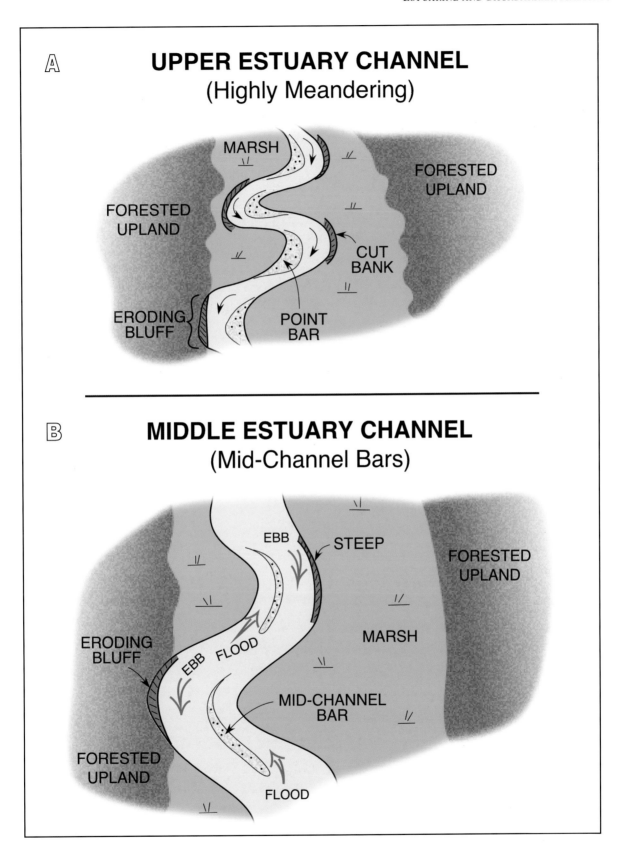

FIGURE 81. General model of the morphology of tidal channels in the estuaries of South Carolina. (A) Upper estuary. (B) Middle estuary.

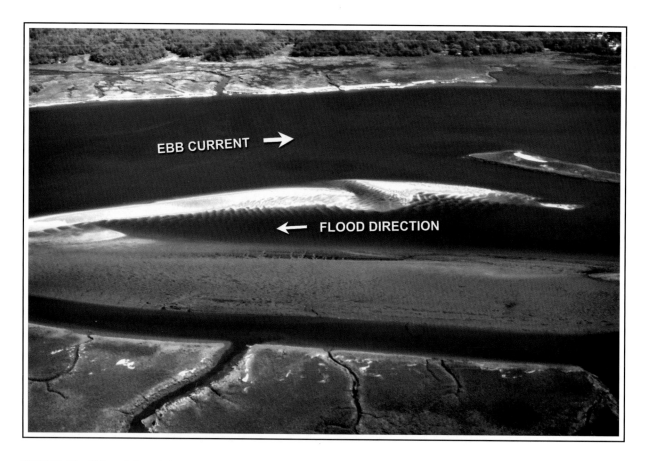

FIGURE 82. Oblique infrared photograph at low tide of a tidal river point bar located on the Wadmalaw River south of Charleston. The ebb-tidal current flows from left to right (arrow). Elevated channel on side of sand spit away from the main channel is dominated by flood currents (arrow; note flood-oriented sand waves in channel; arrow points in direction of the flood current at high-water level). Compare this image with the diagram in Figure 81B. The gray sediment between the marsh and the flood channel (lower half of image) is soft mud. Photograph taken circa 1982.

to be safe from all but the most extreme hurricane. However, a channel with a cut bank that incises into the forested upland adjacent to the channel is a cause for concern in some localities. A photograph of one such eroding bluff is shown in Figure 83. In a number of localities, shore protection structures, such as riprap revetments, have been built to slow down the erosion.

As some developers consider this type of erosion problem, two questions related to the understanding of the mechanisms of erosion of the bluffs are: 1) When do the bluffs erode? and 2) What processes may cause them to erode at rates faster than that shown in the past few decades? As already noted, the bluff erodes when the meandering channel

migrates into it. However, this erosion can only take place when the water is high enough to impact the eroding scarp on the bluff for a significant amount of time. This can happen in the following ways:

1) **During maximum spring tides**, especially the extraordinarily high tides that typically occur during the late fall. It is unlikely that any significant erosion occurs during neap tides.

2) **During storm surges** that may occur during tropical storms or hurricanes. Elevated tide levels also occur during periods of strong onshore winds not associated with tropical depressions, but they would only be a factor during maximum spring tides.

FIGURE 83. An eroding bluff on the shoreline of the May River, South Carolina. The location of this bluff is indicated by the arrow on the image in Figure 80. Photograph taken around mid tide on 30 October 2003.

3) **During periods of high freshwater runoff.** Even in estuaries with a relatively small freshwater stream entering it, periods of very high rainfall may significantly raise the water level in the estuary, creating very strong ebb currents that could erode the bluffs. Once while investigating such an issue, we downloaded the U.S. Geological Survey discharge data for the Coosawatchee River. These data showed that there were some high flows in 2003, but a much more significant high occurred in 1998 when *El Niño* rains drenched the southeast. We had noted a similar trend for the Congaree River. Therefore, more erosion of bluffs in the upper reaches of some estuaries can be expected during the *El Niño* years or possibly during high runoff generated by rains associated with tropical cyclones.

4) **During periods of high rainfall in the local area.** The soil along the banks of the channel may become saturated during periods of heavy rainfall causing slope failure (Tim Kana, pers. comm.).

An obvious potential cause for an increasing erosion rate of the bluffs would be an abrupt rise in sea level. As noted earlier, sea level is rising at the present time at the rate of around one foot per century. If sea level continues to rise at its present rate, the amount of bluff erosion should not increase significantly as a result of that factor alone (at least not for the next few decades).

121

8 THE FRESHWATER RIVER SYSTEMS

DESCRIPTION

Considering the freshwater rivers of the southeastern region of the United States as a whole, they can be classified into three fundamental types on the basis of regional physiography and water source/chemistry: 1) those streams that originate in and derive most of their waters from the Blue Ridge Mountain and Piedmont Physiographic Provinces (here termed **piedmont rivers**; also known as "red-water rivers"); 2) those streams that have most or all of their drainage basins located within the Coastal Plain Physiographic Province (here termed **coastal plain rivers;** also known as "black-water rivers"); and 3) the **spring-fed rivers** of the Florida Peninsula, which clearly require no further discussion in this book.

There are eight primary piedmont rivers and twenty two fairly large coastal plain rivers in the Georgia Bight region (see Figure 84). Three piedmont rivers, the Pee Dee, Santee and Savannah Rivers, deliver sediments to the South Carolina coast. The largest of the coastal plain rivers in the state include the Waccamaw, Black, and Edisto Rivers. There are several fundamental differences between these two types of rivers. For example, piedmont rivers (within the Piedmont Physiographic Province) typically have steeper gradients, more flashy discharge, and less abundant

associated wetlands than coastal plain rivers.

As shown in Figures 1 and 85, the boundary between the Piedmont Physiographic Province and the Coastal Plain Physiographic Province is marked by the presence of what is known as the **fall line**. At that point, the waters of the stream cut down into the older igneous and metamorphic rocks of the piedmont and usually form a zone of rapids. Seaward of the fall line, the river typically flows over a bed of unconsolidated sediments. During the Pleistocene glaciations, when sea level was hundreds of feet below its present level, the zone where the stream was cutting down into the older bedrock extended much further seaward than it does today. Accordingly, as noted earlier, a valley as much as a hundred feet deep was cut across the coastal plain and the present continental shelf as a result of the lowered ocean level. When sea level started to rise again, the carved valley started to aggrade (fill up) with sediments. Ever since the beginning of the aggradational phase, the rivers in the coastal plain area have migrated back and forth across the old valley, filling it in evenly with sediments. The sediments that have filled in the valley have created what is known as the **flood plain** of the river. The flood plains of the large piedmont rivers that cross the coastal plain of the southeastern United States are typically quite wide, several miles in some instances. The general model

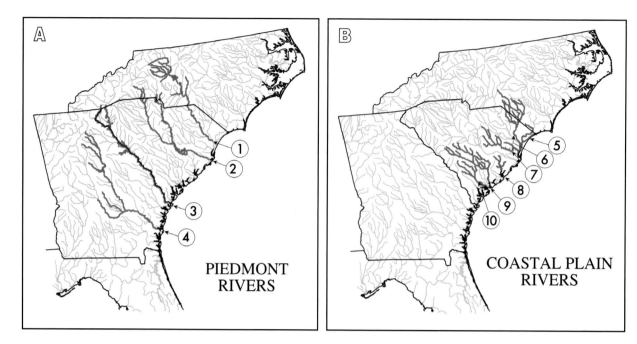

FIGURE 84. Major piedmont and coastal plain river systems occurring within the Georgia Bight region. (A) Piedmont Rivers – Pee Dee (1), Santee (2), Savannah (3), and Altamaha (4). (B) Coastal Plain Rivers – Waccamaw (5), Little Pee Dee (6), Black (7), Cooper (8), Edisto (9), and Combahee (10). Coastal plain rivers in Georgia are not named.

for a flood plain of a typical piedmont river in the Coastal Plain Physiographic Province is given in Figure 86. The flood plains of the larger coastal plain rivers may also be more than a mile across. Because of the differences in bedrock erodability and topographic variability, some rivers in the Piedmont Physiographic Province also have minor flood plains, but these narrow flood plains are invariably separated by a reach of rapids located in confined, erosional valleys.

Hydrographic data obtained from the U.S. Geological Survey are highly detailed and complex but much of relevance is to be learned from them. In a recent study, we concentrated on three rivers in South Carolina with which we are very familiar and where we conducted some preliminary field work: 1) South Fork of the Edisto River, the largest coastal plain river in the Georgia Bight; 2) Broad River, the main stem of the Santee River drainage basin; and 3) Tyger River, a moderate sized piedmont tributary of the Broad. The hydrographs for four separate months during 1991 for these three rivers

are shown in Figure 87. Note that the discharge of the two piedmont rivers is quite flashy, with severe flood peaks occurring several times during 1991, but that the South Edisto discharge was quite steady.

The contrast in the hydrographs of the two river types is determined by the different rates of infiltration and on-land water storage during and after rainfall. Generally speaking, the soils of the coastal plain, particularly the upper coastal plain, are sandy and quite permeable (i.e., they have a high infiltration capacity), whereas the lateritic clayey soils of the piedmont are not. Another factor is the abundance of swamps and bottomland hardwoods scattered throughout the coastal plain, which act as large "storage tanks" for the runoff from the rains. Extensive swamps are absent from the flood plains and drainage basins of the piedmont. These two factors combined promote rapid runoff in the piedmont and slow runoff in the coastal plain, hence the striking contrast in the hydrographs presented in Figure 87.

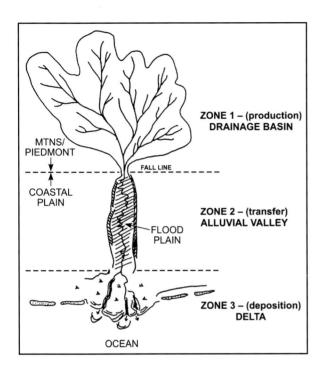

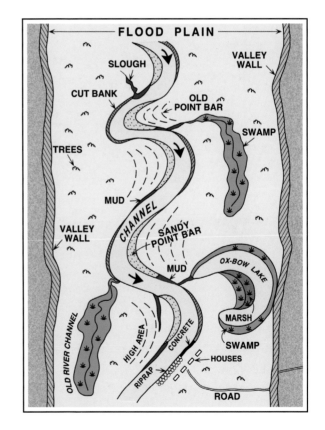

FIGURE 85. The three zones of the direct transfer of sediments from the uplands to the coast. Zone 1 – The watershed or drainage basin located in the Piedmont and Blue Ridge Physiographic Provinces. Zone 2 – The alluvial valley that cuts across the Coastal Plain Physiographic Province. Zone 3 – The coastal zone where most of the sediments accumulate in the deltas and estuaries. Note that the fall line is the boundary between Zones 1 and 2. The physiographic provinces in South Carolina are mapped in Figure 6.

BOTTOMLAND HARDWOODS

Freshwater marshes are rare in the piedmont and coastal plain river systems in South Carolina, the dominant wetland type being **bottomland hardwood ecosystems**. Clark and Benforado (1981), Mitsch and Gosselink (2000), and Taylor et al. (1990) described the zonation of the bottomland hardwood ecosystems of the southeastern United States relative to the main channel of the associated stream. Six zones were described, ranging from Zone I, which is the permanently wet stream or river itself, to Zone VI, which is the transition zone between the floodplain and the uplands and is rarely flooded.

These zones have very distinct plant assemblages. Zones II and III of the bottomland hardwoods are the most frequently flooded (more than 50% of the years of record and typically >25% duration during the growing season), and Zones IV and V considerably less so. For purposes of discussion in this book, we have combined Zones II and III into one class, referred to as **cypress-tupelo swamps,** because two readily recognizable tree species, bald cypress (*Taxodium distichum*) and water tupelo (*Nyssa aquatica*), are the predominant vegetation of the two classes and their ecological conditions are quite similar. Zones IV and V are also lumped into a single class, **upper bottomland hardwood forests,** which have a much more diverse population of plant species. A generalized flood plain cross-section showing the different

FIGURE 86. Plan view sketch of the morphology of the flood plain of a typical major piedmont river where it crosses the Coastal Plain of South Carolina. These flood plains are typically several miles across.

125

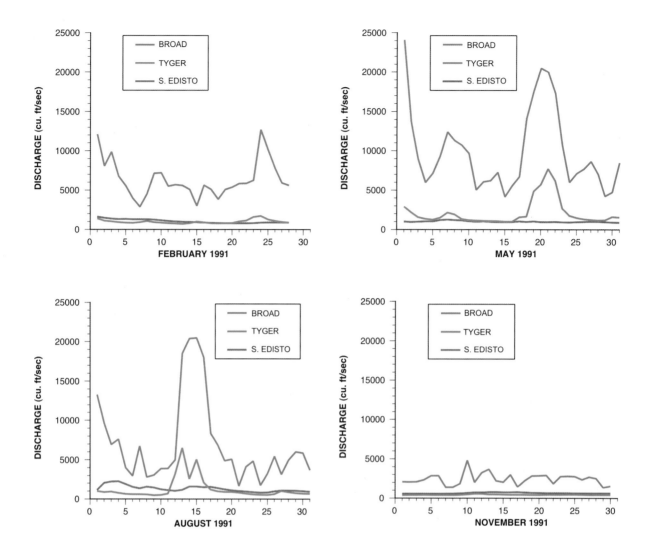

FIGURE 87. Hydrographs for two piedmont rivers (Broad and Tyger) and one coastal plain river (South Edisto) in South Carolina during four separate months in 1991. Note the flashy discharge of the piedmont rivers compared with the relative steady discharge of the coastal plain river during the same time frame.

zones of the bottomland hardwood system as we have combined them, as well as descriptions of the wetting conditions and lists of the most common trees in each of the redefined zones, is shown in Figure 88. Photographs representative of the two primary zones are given in Figure 89.

Our observations of piedmont and coastal plain rivers in South Carolina revealed striking differences between the flood plains of the two river types. Typically, piedmont rivers in the Piedmont Physiographic Province have high, narrow flood plains, several feet above the mean water level of

the stream. No cypress/tupelo swamps are present. Upper bottomland hardwood forests are present in places but not common. The sediments are usually sandy. Coastal plain rivers, on the other hand, have low, relatively wide flood plains, commonly only inches above the mean water level of the stream, and cypress-tupelo swamps are present in extensive areas across the flood plain.

We believe the high flood plains of the piedmont rivers in the Piedmont Physiographic Province are the result of the flashy discharge of the streams. After heavy rains, the river rises quickly

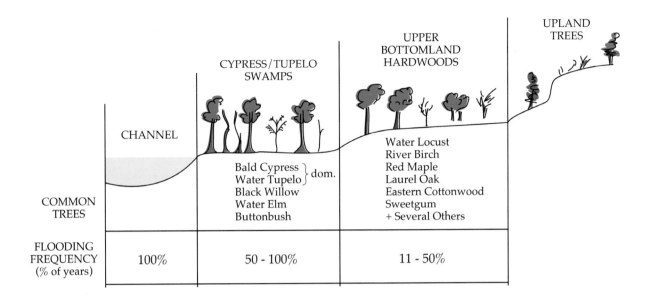

FIGURE 88. Topographic cross-section illustrating the components of typical bottomland hardwood ecosystems in the southeastern United States. Modified after Clark and Benforado (1981), Mitsch and Gosselink (1986), and Taylor et al. (1990).

to a high level and sediments are transported up to the surface of the flood plain, building it up because of both the extreme water level and the large transporting capacity the stream has for sediments. As a general rule, the coastal plain rivers have a steady discharge minus the ultra high water levels, and the flood plains are not built up as high as those on the piedmont rivers upstream of the fall line. With the low banks on the main channel, these rivers frequently "leak" into the adjacent cypress-tupelo swamps.

Where the three piedmont rivers present in the state – Pee Dee, Santee, and Savannah Rivers – cross the coastal plain of South Carolina, they are characterized by: 1) flood plains several miles wide; 2) multiple abandoned channels which contain oxbow lakes in places; 3) extensive cypress-tupelo swamps where the abandoned channels have been partially filled with sediments; 4) very extensive upper bottomland hardwood forests, some with old growth trees (e.g., in Congaree National Park); and 5) meandering channels with abundant sandy point bars along the outer bends of the meanders (Figure

86). Coastal plain rivers have similar, but smaller scale, features in their flood plains.

FIGURE 89. Examples of bottomland hardwood ecosystems. (A) Upper bottomland hardwoods. Note the abundance of dwarf palmetto (*Sabal minor*) in the understory. (B) Cypress/tupelo swamp. The closest tree in the center of the photo is a water tupelo (*Nyssa aquatica*) and the large tree in the middle distance is a bald cypress (*Taxodium distichum*). Note the abundant cypress knees around the cypress tree.

9 CONTINENTAL SHELF

Though not a place you are apt to visit during your naturalist explorations on the South Carolina coast, unless you own or hire an ocean-going vessel, the **continental shelf** represents the end member of the coastal zone continuum. The shelf off South Carolina is quite wide, extending an average of about 60 miles off the beach. A distinct increase in slope at depths of around 600 feet marks the edge of the shelf and the beginning of the continental slope. Figures 8 and 90 show some details of the bottom topography of the shelf. It is clear from these illustrations that the continental shelf is not a boring flat surface but has an abundance of topographic features, including:

1) There is a triangular bulge projecting off the continental slope margin that is located directly offshore from the mouth of the Santee River. That promontory is most likely what geologists call a **lowstand delta**. During one of the major glaciations, sea level was low enough for the mouth of the river to spill its sediment over the edge of the slope forming the prominent protuberance at the shelf edge. This particular delta has not been studied, but several ones at the edge of the shelf in the Gulf Coast region have (e.g., Sydow and Roberts, 1994).

2) Several shoal complexes exist on the shelf closer in to shore in the vicinity of the present mouth of the Santee River (Figure 90). According to research carried out by one of our former colleagues, Walter J. Sexton, who is now the president of Athena Technologies, Inc. in Columbia, the shoal directly off the river mouth is an ancient delta deposit formed during a stillstand episode when sea level was rising after the last glaciation (Wisconsin). The lobe off Bulls Bay was a delta lobe deposited during an earlier rise of sea level after one of the older glaciation episodes. The shoal directly off Cape Romain may have been formed as a barrier island complex related to the formation of an ancestral cape, again during a pause in the rising sea level, in this case, probably the last one.

3) The bathymetric data of Figure 8 also shows the presence of gigantic ebb-tidal deltas, as for example, the one directly off the entrance to Port Royal Sound.

These features indicate that the post-glaciation rise of sea level was anything but steady, with a number of pauses, or stillstands, taking place from time to time.

Another consequence of the sea-level fluctuations during the Ice Ages was the carving of valleys (**lowstand valleys**) across the continental shelf when sea level was at its lowest levels. A

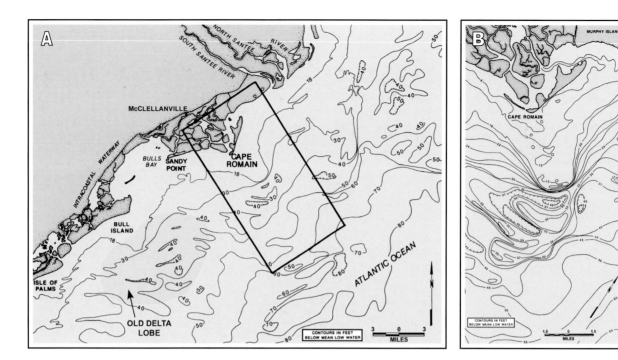

FIGURE 90. Bathymetry of the continental shelf off the central coast of South Carolina. (A) This complex offshore bathymetry is thought to be the result of changes of the position of the mouths of the Santee and Pee Dee Rivers at different times and stages of sea level. The old delta lobe seaward of Bulls Bay is particularly significant, as we believe it to be the source of the sands of the barrier islands to the southwest. The rectangle outlines the shelf area studied in detail by Sexton (1987) and shown in diagram B. (B) Detailed bathymetric map of the seafloor in the vicinity of the Cape Romain shoal. Note the steep scarp at the seaward end of the shoal.

detailed study sponsored by the U.S. Geological Survey showed that seven major ancient lowstand valleys occur on the inner shelf in the area from the present mouth of the Pee Dee/Waccamaw rivers to the North Carolina border, a distance of approximately 56 miles (Baldwin et al., 2004; discussed further in Section II). When sea level was rising, these valleys were filled with sandy sediments in places as the streams built their flood plains within the valleys. Such sandy deposits have been dredged and used in a number of beach nourishment projects in the Myrtle Beach area. These nourishment projects will be treated in more detail in Section II under the discussion of the Grand Strand compartment.

10 POTENTIAL RECREATIONAL ACTIVITIES

CANOEING AND KAYAKING

We have floated our canoe along many of the streams in the coastal plain of South Carolina, and our favorite ones to float are the black-water rivers. The water levels of these streams are pretty consistent; they do get high at times but not as high and muddy as the piedmont streams. They mostly have highly vegetated and isolated stream banks, with little development along them. The ones we have floated the most are the Edisto, the Black, and the Lynches. The Little Pee Dee River, which has some exquisite preserved wetlands along its banks, has engendered rave reviews by others, but we have never floated it. Our favorite time to float is in early spring. Late March is good. The flowers are beginning to bloom and the summer birds are coming back and singing away. The book, *Paddling South Carolina*, revised in 2001 by Gene Able and Jack Horan, provides information on 30 of the most floatable streams in the state.

On the coast, there are many places to launch a kayak (or a canoe) in order to enjoy the tidal creeks and marshes. There are some professional kayaking outfitters and guides on the coast that sponsor tours.

BIRDING

Of course, inasmuch as we are ardent birders, we think the birding is great in South Carolina. On the coastal plain, the key areas not to miss are: 1) Congaree National Park and Francis Beidler Forest for swamp species; 2) Santee Delta area for a mix of swamp and salt-water birds; 3) Pitt Street Bridge in Mount Pleasant for the marsh inhabitants; 4) Throughout the ACE Basin for a variety of species; and 5) Savannah National Wildlife Refuge for ducks! In the discussion of the different coastal compartments that follows, we will point out other areas where we have had good birding outings.

FISHING

We also like to fish, but Michel is too much in love with trout fishing (with artificial flies) in the mountains to spend much time fishing at the coast. On the other hand, Hayes, who also fishes for trout, spends a lot of time chasing largemouth bass in the world-famous bass lakes of the Santee-Cooper group (particularly Lake Marion) and Lake Murray. The Congaree River also has excellent bass fishing. We haven't tried them much, but the coastal black-water streams, such as the Combahee and Edisto, are reputed to be good fishing for bass and sunfish. In fact, there is good freshwater fishing all over the state.

The coastal estuaries have first-class fishing for the very tasty sea trout and flounder. The favorite target of many coastal fishermen is the redfish, which the locals call spot-tail bass. If you like fly fishing, there are several fly-fishing guides in the

Charleston and Beaufort area that can get you onto some spot tails. And, of course, there are also a number of professional offshore fishing charters available.

SHELL COLLECTION

There are a few beaches on the coast where the shell collecting is pretty good. The Edisto Beach State Park comes to mind. Most of the best sites are located in places where you need a boat to reach them. We will point out some of the best places to look in the discussion of the individual coastal compartments.

RELAXING AT THE BEACH

The wide, flat South Carolina beaches are probably the most beautiful on the east coast. Of course, you could rent a hotel room and get to the highly populated beaches at Myrtle Beach. The Myrtle Beach State Park has an attractive beach, but it can be crowded at times. Instead, you might want to seek out some of the less populated and still more-or-less natural beaches in the state and county parks. We think the most rewarding of the more natural sites to visit are Huntington Beach State Park, Folly Beach County Park, Beachwalker Park on Kiawah Island, Edisto Beach State Park, and Hunting Island State Park.

SECTION II
Major Compartments

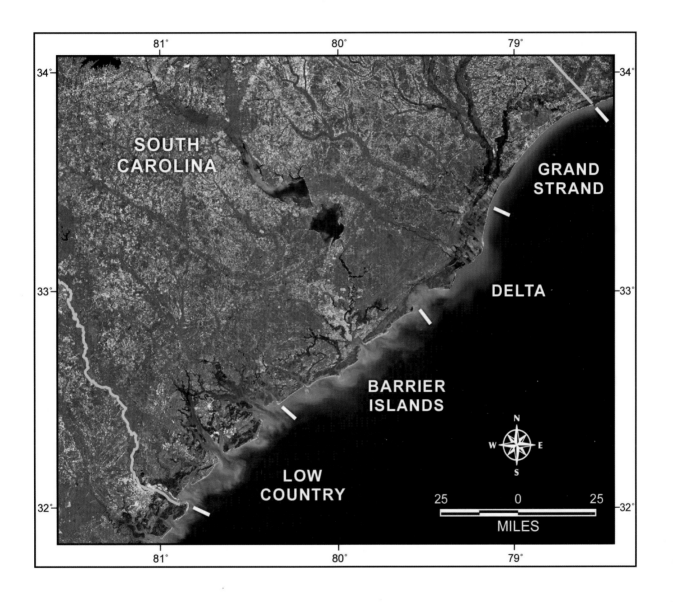

11 THE GRAND STRAND

INTRODUCTION

As you begin your visit, whether starting in this compartment or elsewhere, please be forewarned that there are a large number of gated communities along the coast, some of which provide housing for tourists. Some may also be entered for certain specific activities, such as playing golf. A few even have daily visitation permits (for a fee). If interested in a particular location, visit their web site for details.

In any event, the compartment we call the **Grand Strand** is a **symmetrical shoreline arc** that extends for about 50 miles between the North Carolina border, at Little River Inlet, and Pawleys Island (see vertical image in Figure 91). This is by far the most populated stretch of shoreline in the state. The city of Myrtle Beach is known for its tourism activities, with abundant beach hotels and entertainment centers of different kinds crowded along the sandy beach. The area is also noted as a golfing center, having 40 golf courses and counting within a few miles of the city. Probably the most noteworthy places for naturalists to visit on the open ocean beach in this compartment are the Myrtle Beach and Huntington Beach State Parks.

COASTAL GEOLOGY

Throughout the northern two thirds of this shoreline arc, the modern beach abuts against a formerly eroding Pleistocene upland composed of a series of beach ridges of the Myrtle Beach Formation. Unlike most of the rest of the coastal area of the state, barrier islands are limited to the southern one fourth and the extreme northern part of the compartment. Therefore, there are some small inlet/estuarine systems at the extreme north and south ends of the compartment, but the middle zone has only a few small drainage channels (locally called "swashes") emptying from the upland onto the beach. This absence of the usual barrier island/ backbarrier habitats in this area is probably due to uplift on the Cape Fear Arch (see Figure 4). This gently crescentic portion of the coast has been uplifting and at least slightly erosional from the late Pleistocene to the present.

As reported in a Masters thesis in 1982 by one of Hayes' students at the University of South Carolina, Michael Svetlichny, under normal conditions, the sandy beach sediment had been mainly derived from updrift sources to the northeast and from the fairly resistant Pleistocene barrier deposits, inasmuch as there is presently no direct source of river-derived sediments within this entire compartment. At the time of Svetlichny's study, the outcropping of a semi-consolidated shell rich layer at two localities within the lower intertidal zone, as well as two fishing piers, combined to act as partial sediment traps for longshore sediment transport. Also, toward the northern end of the

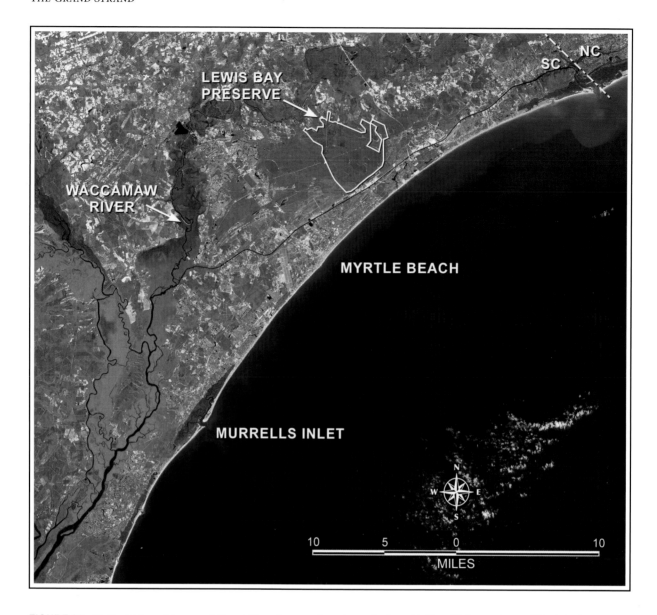

FIGURE 91. Vertical infrared image of Grand Strand compartment acquired on 11 May 2000, courtesy of Earth Science Data Interface at the Global Land Cover Facility. Note that this compartment consists mostly of a narrow beach abutted against the Pleistocene upland, with barrier island systems present only near the SC/NC border and at Murrells Inlet.

beach, sporadic outcrops of mud and peat deposits were exposed during low tide.

More recently, a number of studies have been carried out in the area by the coastal geologists at Coastal Carolina University, located in Conway about 15 miles northwest of Myrtle Beach. Much of this work, some of which is supported by the U.S. Geological Survey, has been carried out by Professors Paul Gayes and Doug Nelson. A study

of considerable interest to us was a survey of the continental shelf of the area soon after hurricane *Hugo* passed, which showed evidence of significant offshore transport of sediments and bottom scour by seaward directed currents as a result of the storm (Gayes, 1991). This no doubt contributed to some extent to the beach erosion that continues to plague some parts of the Grand Strand.

Other studies by the Coastal Carolina group

and their associates at the U.S. Geological Survey have focused on other aspects of the character of the continental shelf off Myrtle Beach as well (Donovan-Ealy et al., 1993; Gayes et al., 2003). Among other things, their results show that the inner shelf is comprised of a patchy, discontinuous sand sheet overlying an erosional surface created as the sea migrated shoreward. Sediment thickness increases to the south, where the largest accumulations of sediment generally occur near tidal inlet systems.

Another very interesting study by this group (Baldwin et al., 2006) dealt in detail with something they called "paleochannel groups" that extend from the mainland out onto the continental shelf. These features are, at least in part, what we described as lowstand valleys in earlier discussions. Some of the most important observations in that publication include:

- Seven major "paleochannel groups" occur in this area from the present mouth of the Pee Dee/Waccamaw rivers to the North Carolina border, a distance of approximately 56 miles.

- Over time, from the late Pliocene to the present (last 3 million years or so), the valleys have been displaced to the south. Two possible reasons are suggested in the paper: a) forcing the stream outlet to the south as a result of southerly longshore sediment transport; and b) continuous elevation of the Cape Fear Arch, which is located a few 10s of miles north of the South Carolina border (Figure 4). The authors of the paper favor option (a), but we favor option (b) because of evidence of the impact of the rising arch we have observed on flood plain geomorphology of several major rivers to the south of the arch.

- Generally speaking, the "paleochannel groups" range from 18-72 feet deep and have widths between 0.6 to 7.2 miles. They are all filled to the top with sediments at the present time.

These valley fills are exposed at the surface

(i.e., of the continental shelf) in places and buried by modern day sedimentation in others. These kinds of studies are relevant to the fate of the local beaches because they help locate sand bodies that can be dredged in the future for sand to be used in beach nourishment projects, a subject we now somewhat reluctantly broach.

THE GREAT BEACH NOURISHMENT DEBATE

The deposition of sand derived from elsewhere onto the intertidal beaches of the South Carolina coast during beach nourishment activities has been taking place for many years, including 15 projects in the 1990s alone. Kana (1990) summarized nourishment activities up to 1990 and predicted future needs for the coming decade and beyond.

Beach nourishment is one of the most commonly used "soft" engineering solutions to beach erosion problems. Studies by our group in the 1970s showed that erosion rates in the Myrtle Beach area were relatively mild compared to some other sections of the coast. However, intense building close to the high-tide line plus the installation of hard structures on the backbeach exacerbated the erosion of the sand beach to the point that the local powers that be requested help to solve their beach erosion "problem." Consequently, considerable effort has been expended to deal with the erosion along these very popular beaches, with beach nourishment being a prime example. We will point out some of the other beach nourishment projects that have been carried out in the state as we work our way down the shore in this discussion of the coastal compartments. Myrtle Beach, one of the first beaches to be nourished in the state, is located in this compartment, thus we will discuss it in some detail here so as to acquaint you with the issues surrounding this technique. Other projects throughout the state will not be discussed in such detail, because the Myrtle Beach example adequately covers the subject, in our opinion.

The history of the beach nourishment efforts at Myrtle Beach has led to a lively debate, which was published in the South Carolina Business & Economic Review. It all started with an article published in the summer of 2006 by Dr. Tim Kana, our former partner at RPI, who is now President of Coastal Science and Engineering, LLC, a company located in Columbia that has probably done more beach nourishment projects in the state than any other group. The other participants in this debate are our old buddy Dr. Orrin H. Pilkey, emeritus professor at Duke University, and the combined efforts of Drs. Jeff Pompe and Jim Rhinehart, professors at Francis Marion University in Florence, South Carolina, who later published responses to Kana's article in the same business journal. We can't help but call this debate the GOOD, the BAD, and the UGLY, because of the extreme divergence of opinions expressed. Soon after writing this section, we discovered that Pilkey and co-author had recently used the good-bad-ugly epithet in a paper on the effects of beach nourishment on birds. Sorry, Orrin, we can't go back now.

In any event, we believe this discussion about nourishing Myrtle Beach puts the whole issue clearly before us and probably applies equally well to such projects just about anywhere else, most assuredly anywhere on the coast of South Carolina. After reviewing the different arguments presented in the business journal, which are summarized below, hopefully, we will all be better educated on this subject.

The following summary points are taken directly from each of the three papers. Our comments are shown in italics. Let us begin with the different aspects of the projects at Myrtle Beach from Kana's viewpoint:

THE GOOD

1) In the early 1980s, beach erosion was bad, with little beach at high tide. (*Not as bad as a lot of other areas in the state, but definitely one that*

had the mayor worried.)

2) Revetments, seawalls, and sand bags protected buildings, parking lots, and swimming pools from damaging surf.

3) He outlined the different nourishment projects carried out in the area up to *Hugo* (September, 1989), and was confident the nourished sand that had been placed on the beach earlier was a great benefit to the beach, because it diminished potential erosion during that hurricane. (*Definitely less erosion than elsewhere, as we saw during some of our post-storm surveys.*)

4) The long-awaited federal nourishment project was finally completed in 1997. The U.S. Army Corps of Engineers restored Myrtle Beach using offshore sand pumped into place by dredge. At nearly double the size of the local project (done previously), the 1997 nourishment buried seawalls, reestablished a dune line, and widened the beach by about 100 feet. Total cost of the project was $15 million. More than 70 percent of the sand placed by the Corps remains on the beach today. (*Remember, this was written in the summer of 2006.*)

5) Restoration of Myrtle Beach's eight-mile strand has cost an average of only $20 per foot of beach per year since 1986. (*We haven't checked it, but we believe his arithmetic.*)

6) How Myrtle Beach went from being a poster town for beach erosion to a place where few even worry about an encroaching sea is a lesson for other beach communities. (*Happy news for some coastal engineers, as well as beachfront property owners.*)

7) People who back in the early 80s paid $10,000 per foot of shoreline for a hotel site, even without a beach at high tide, now sit on property worth as much as $100,000 per foot of beach. (*Darn! And here we were buying plantations in the early 1980s when we should have been buying lots and/ or hotels on the beachfront at Myrtle Beach.*)

8) Their lots didn't wash away, their seawalls didn't fail, and they now look out over a band of dunes

to a white, sandy beach that accommodates thousands of visitors each day—even at high tide. (*In the old days, there was little dry sand beach at high tide.*)

Now, let's continue with Pilkey's viewpoint:

THE BAD

1) *Hugo* damaged the other beaches in the area more than those at Myrtle Beach because they are narrow and much lower-elevation communities with little dune development. (*There is some truth to this, especially at Pawleys Island, as you will see in later discussion.*)

2) A big storm like *Katrina* would play havoc with this beach development, nourishment or no nourishment. (*No doubt about it! We are experts on this subject. We were living in New Orleans when Katrina hit.*)

3) Beach nourishment is very damaging to the beach biota (i.e., the combined flora and fauna of a region) where a complete and unique food chain exists, topped by the birds people love to watch and the fish people love to catch. No beach biota—no fish and no birds. (*Kana says we are doing a better job at this than we used to. However, some prominent biologists in our acquaintance agree with Pilkey on this one, well, at least to the fact that these projects do harm to the biota. We are not so sure about the no fish part. So you might want to go ahead and bring your fishing rods when you visit.*)

4) Dredging sand from offshore harms the biota there as well. Communities like Myrtle Beach have turned the nearshore zone with its precious ecosystem into an engineering project. (*As we said before, Pilkey has this thing about engineers.*)

5) Beach nourishment encourages high-density development immediately adjacent to the beach. In a time of rising sea level at a location where big storms are a certainty, beachfront construction is at least irresponsible and at

worst can only be considered a form of societal madness. (*Sometimes you are not very sure just how Pilkey really feels about this subject.*)

6) Why should federal or state taxpayers pay for nourishment if it will destroy the aesthetics of a beautiful beach and encourage increased density of development? [*An argument we bet the readers of this book will find difficult to refute. However, some beach nourishment projects are paid for by the folks who actually live at the beach (e.g., beach owners associations, town governments, etc.), aesthetics destroyed or not.*]

7) The most fundamental problem with the high-density development encouraged by beach nourishment is that it allows no flexibility for future response to sea-level rise. [*Water usually runs downhill, too. And the sun will probably rise tomorrow, etc., etc. Wonder what flexibility the plans for New York, Boston, and Miami have? What about Charleston? Not to make light of this point, but, heck, a significant percentage of our lawmakers don't even "believe" in global warming. Not to mention the letter to the editor in today's paper (The State newspaper in Columbia), which stated that our "God-given common sense" clearly tells us that global warming is a bunch of hokum. We are not sure where they (the politicians and the letter writer) get their science from. Oops, just as Hayes was writing that last statement a couple of weeks ago, he glanced at that day's copy of the The State newspaper and lo and behold there was a headline regarding our recently re-elected governor, Mark Sanford. Anyway, the headline read – SANFORD WANTS S.C. TO LEAD GLOBAL WARMING FIGHT. Wow!*]

8) The national bill for nourishment is already more than a billion dollars per decade. However, the federal government has effectively dropped out of the program and public support is waning because of the growing perception that beach nourishment is a form of welfare for

the rich. (*Some of whom are engineers doing this kind of stuff.*)

We conclude with some of the comments by the two professors from Francis Marion University.

THE UGLY

1) Much of the 198 miles of South Carolina shoreline is covered with wide, sandy beaches that are responsible for a significant portion of the $14.6 billion tourism industry in the state. Without high-quality beaches, many of the 28 million visitors who flock to the state each year would take their dollars to other East Coast beaches. (*Hmmm. That many miles of wide sandy beaches? Okay, are they washing away or are they not washing away? And you guys thought you came to the coast to look at birds.*)

2) As one might expect, this population growth has increased the demand for coastal property, thus pushing up housing prices significantly. (*Darn again! Boy, we wish we still had that plantation.*)

3) Beaches provide recreational benefits for residents in coastal areas as well, where population has increased dramatically in recent decades. One million people now live in the state's coastal counties, and the region has four of the five fastest-growing counties in the state. (*Who wouldn't want to live on this beautiful coastline?*)

4) At first glance, it might look foolish to replace sand that will simply wash away with the next big storm, but following this line of thought, it also looks foolish to build a house that rots, peels, and decays when subjected to wind, rain, hail, heat, cold, and insects. (*Take that, Pilkey!*)

5) We believe that better and fairer decisions about beach restoration could be made if a payment schedule were designed that would link benefits accruing to certain individuals with taxes and fees paid by those same individuals. (*Oops.*

Guess it is a good thing we didn't buy that hotel after all.)

6) A referendum could be held whereby local taxpayers would vote on beach nourishment proposals, thereby levying taxes on themselves for the projects from which they derive value. (*Don't these guys know they live in a Republican state? Raise taxes hell!*)

7) At least one South Carolina community has attempted to consider a better balance between those who benefit and those who pay. On DeBordieu Island, residents are responsible for all nourishment costs because the community is gated and the beaches are not open to the public. For a recent $2 million nourishment project, beachfront owners were required to pay a higher fee than those not on the oceanfront. (*Bet those guys could afford it, though.*)

8) Protecting the South Carolina shoreline in the future may become more costly than our history would suggest. Over the past 80 years, the sea level in Charleston has risen 10 inches, etc., etc., etc. (*After listening to all this, we think we will just stay in our mountain place, go trout fishing, and forget the coast.*)

TO DO OR NOT TO DO

We made some tongue-in-cheek remarks in that last section, but we don't want to disparage in any way the participants in the debate. We do not personally know the two writers of the last paper, but we have a long history of association with the other two. Tim Kana did his dissertation at the University of South Carolina under Hayes' supervision, and in 1976, they co-authored a book, *Terrigenous Clastic Depositional Environments,* which presented some of the earliest data on the physical makeup of the South Carolina coast. Also, before he formed his present company, Tim was our partner at RPI for several years. Tim provided to us many of the details on the recent work on the coast that we cite in this book. As professional geologists and members of the same professional societies, as well as through University contacts, we have had many interactions with Orrin Pilkey over the years.

Kana and Pilkey, who both have an abiding love and respect for the natural environment, obviously approach the subject in a different way. Kana has a practical approach, which we certainly sympathize with, and he strives to provide practical solutions to problems presented by his clients whenever possible. When RPI followed up on a suggestion our group made to relocate Captain Sams Inlet in 1983, Tim Kana was the on-the-ground supervisor of the project. He also continued to monitor the project in later years and has published several papers on the results (e.g., Kana and McKee, 2003).

Obviously, Pilkey is different, preferring a "let it be" philosophy. He has an agenda, which we also sympathize with, and he sticks to it, much to the chagrin of many coastal engineers who question the authenticity of some of his data and suppositions. The clearest exposition of his philosophy we can find in his writings appear in the final chapter of his artfully illustrated and informative book, *A Celebration of Barrier Islands,* published in 2005. In this book, he suggests that barrier islands lead a Gaia-like existence. As you probably know, the Gaia hypothesis is an ecological theory that proposes that the living matter of planet Earth functions as a single organism. Beliefs similar to this go back at least to the great Spinoza in the mid 1600s, whose philosophy was cited, but not necessarily believed, by Einstein. Actually, this concept goes back thousands of years, but you probably get the point.

To quote Pilkey's summary of this matter exactly, he said:

"Putting barrier islands in a Gaia context does provide a basis for thinking about how to live with them in a way that will keep islands existing and evolving, a way to preserve them for future generations. Perhaps more important, the distinction between life and death, as it applies to a barrier island, should provide guidance

to distinguish good and bad development practices."

Strangely enough, in co-author Hayes' book entitled *Black Tides,* published by The University of Texas Press in 1999, a sentiment similar to Pilkey's was expressed. In this book, which deals with our experiences responding to most of the major oil spills that have occurred since 1974, reference was made in the last chapter to a short article entitled "Ensoulment of Nature" by Gregory A. Cajete. That paper was published in the book *Native Heritage.* Cajete stated in the article that the importance the American Indians put on connecting with their place of origin (i.e., the environment) is not just a romantic notion out of step with the times, but rather "the quintessential ecological mandate of our times." He stated further that the native peoples experienced nature "as a part of themselves and themselves as a part of nature." Under this concept, **ensoulment** means that the human participates with the earth as if it were a living soul. Hayes concluded this subject with the following comment:

"If we all adopted this frame of mind, I'm sure it would be one more step in the honorable direction toward making the earth a better place to be, which is something all of us tree-huggers and right-thinking engineers want to do, correct?"

That final chapter in *Black Tides* concluded with a quote from the Bureau of American Ethnology Collection (published in Brown, 1970), which follows:

The old men
say
the earth only
endures.
You spoke
truly.
You are right.
The earth only endures.

Of course, the animals and plants that lived on the earth have not fared so well at times. An example of those hard times being when the earth was struck by a large asteroid or comet, such as the one proposed to have caused the Cretaceous/Tertiary extinction event about 65 million years ago. But, sure enough, the earth itself has endured, so far.

With that behind us, we will now describe some of the more pleasant and productive localities for a naturalist to visit in the Grand Strand.

THE NORTHERN BORDER

A splendid drumstick-shaped barrier island almost three miles long, still primarily in its natural state, is located just west of the Little River Inlet, a jettied inlet at the North Carolina/South Carolina border. Just for the record, in 1992, a total of 1,049 acres on the island (Waites Island) and the adjoining uplands of Tilghman Point were donated to the Coastal Education Foundation, Inc. The gift has been protected with a perpetual conservation easement through The Nature Conservancy of South Carolina. Students and faculty at Coastal Carolina University have access for research and other University related activities to a portion of this island. Therefore, access by driving is limited mostly to University students and staff. See vertical image of this area in Figure 92.

MYRTLE BEACH STATE PARK

The last time we paid a visit to Myrtle Beach, in January 2007, we were driving around looking for the entrance to Myrtle Beach State Park. When we found it, Michel was heard to say, "well here it is, right in the middle of the nastiest bunch of foo foo crap there is."

Okay, so you have to drive off US 17 a ways to get back to nature. Some appealing trees (live oaks, magnolias, etc.) line the entrance road, and there is a picturesque pond near the second entry structure.

Along this drive would be a good birding spot for land birds.

Myrtle Beach State Park, which is host to over one million visitors per year (open 6 am to 10 pm daily; entry fee of $4 for adults), was first opened to the public in 1936, becoming South Carolina's first state park. Now, as you will no doubt soon discover for yourself when you visit the area, its 312 acres include some of the last natural area remaining on the coastline of the Grand Strand. A nature trail works its way through a 100-acre stand of maritime forest, one of the last stands in the Grand Strand (a Heritage Trust Site).

This park has a very beautiful mile-long, wide, white sandy beach (yes they do still exist here) still in its pristine natural state. Wait a minute. Tim Kana informs us that this beautiful sand was derived from two beach nourishments next door.

Still it is odd, a surprise even, to see such a delightful beach pinned in between two such highly developed areas. When we were last there in January 2007, a foredune ridge several feet high topped with sea oats (*Uniola*) stretched the entire length of the undeveloped beach. The leading edge of the forest behind the beach exhibited some spectacular **salt pruning** of the wax myrtle and live oaks. Salt pruning of the shrubs and trees occurs along the backbeach area because the constantly breaking waves throw a fine mist of salt water into the air. During landward sea breezes, the salt spray stunts the growth of the plants on the side closest to the beach. The result is that the vegetation looks like it has been pruned shorter in the front and higher in the back, making it appear to bend away from the sea. More than 190 species of birds have been recorded in the park, with peak viewing in the land area being during the spring and fall, when migrating song birds, such as warblers and vireos, put in their appearance.

Other features and potential activities at the park, in addition to swimming, of course, include:

- Broad parking area with associated rest rooms

FIGURE 92. The northernmost portion of the Grand Strand compartment. The undeveloped, drumstick-shaped barrier island located just west of the jetties is Waties Island. The river in the middle of the image (Little River) is the South Carolina/North Carolina border. Infrared image acquired in 2006, courtesy of SCDNR.

and picnic tables

- Accommodations that include nearly 350 campsites, five cabins, and two apartments
- Nature and activity centers, playground, and boardwalks
- Surf fishing

The centerpiece of the park is a long fishing pier, where, after paying a small fee for bait, etc., you may fish for flounder, spot, drum, and blue crabs, among others. Fishing or not, you should take a walk out on the pier (hours 10-5) for an overview of this exceptional beach, the well-developed foredune ridge, breaking waves, and the spectacular salt pruning. Photographs taken from the pier are shown in Figure 93. During our short visit in January 2007, we observed the following birds from the pier: ring-billed gulls, a ruddy turnstone, northern gannets, brown pelicans, cormorants, and terns.

A word of warning: This park is going to be crowded in the summer time.

FIGURE 93. Views from the fishing pier at Myrtle Beach State Park. (A) Southerly view at near low tide. Arrow points to salt-pruned shrubs and trees behind the beach. (B) Northerly view. Photographs taken on 14 January 2007.

LEWIS OCEAN BAY HERITAGE PRESERVE

The coastal area in the Grand Strand is by far the most intensely developed part of the coast. If you want to get away from all this and take a walk in a truly isolated area go to the Lewis Ocean Bay Heritage Preserve (located on Figure 91). This preserve is managed by the South Carolina Department of Natural Resources (SCDNR), which has provided us with the following information about it.

The preserve can be reached from SC 90, which intersects US 501 about two miles south of Conway. From that intersection go approximately 6 miles to the east along SC 90. From there go southeast on International Road for 1.5 miles, where you take a left on Kingston Road. The preserve, which is open during daylight hours year-round, lies on both sides of the Kingston Road. Jamie Dozier, wildlife biologist in charge of Coastal Preserve Management, urges visitors to park their vehicles and walk along the well-maintained roads to experience the flora, fauna, and a wonderful sense of solitude.

"This is one of the few places in Horry County where that's possible (to experience such solitude)," Dozier says. All of the coastal zone of the Grand Strand compartment north of Murrells Inlet is in Horry County (Figure 91).

This huge preserve, which occupies the surface of a highstand Pleistocene barrier island complex called the Jaluco barrier (around 200,000 years old), covers 9,383 acres. It has the largest number of black bears anywhere in the state, except for the mountains. The bears migrate along the Waccamaw River Corridor between North Carolina and this part of South Carolina. Bald eagles and red-cockaded woodpeckers are also residents of the area, as well as numerous songbirds. We have never birded this preserve, but we plan to sometime soon.

Some very interesting plants, such as the rare insectivorous Venus flytraps, pitcher-plants, native orchids, azaleas, and other wildflowers bloom in profusion in season. Seasonal changes are noteworthy, for example:

- January-March – wire grass and migrating songbirds are big attractions
- March-June – pitcher-plants and wildflowers are in bloom
- Fall – songbirds abound and other wildflowers (e.g., blazing star) are in bloom

Inspect the aerial photograph in Figure 94 and you will see a large number of elliptical ponds and wetland areas called **Carolina Bays** scattered throughout the preserve. These types of bays, which range in length from 30 feet to almost 3 miles, are a geomorphic feature present only in the Coastal Plain Province, occurring from northwest Florida to New Jersey. More than 100,000 of these bays are known to exist, and they are best developed in North and South Carolina. The two most striking characteristics of the bays are their elliptical shape and their consistent northwest-southeast orientation. There are 23 of them within the preserve, the largest number of undisturbed Carolina Bays in one place in South Carolina!!

In scientific lore, these "bays" have been ascribed to at least 18 different origins, one of the most popular being meteorite impacts. However, there is no evidence whatsoever of meteorites or meteorite fragments being buried deep beneath these features. More modern studies, including a dissertation at the University of South Carolina by Ray Kaczorowski (a Hayes student), who created similar features in the laboratory, show that their origin is not so exotic (Kaczorowski, 1976). They are now thought, by us at least, to have been formed as lakes carved into their elliptical shape by wind-driven waves and currents back over the years, at times when the ground water levels were high enough for the original lakes to develop. Under each lake, an aquitard (layer through which water will not penetrate) allowed the water to perch above

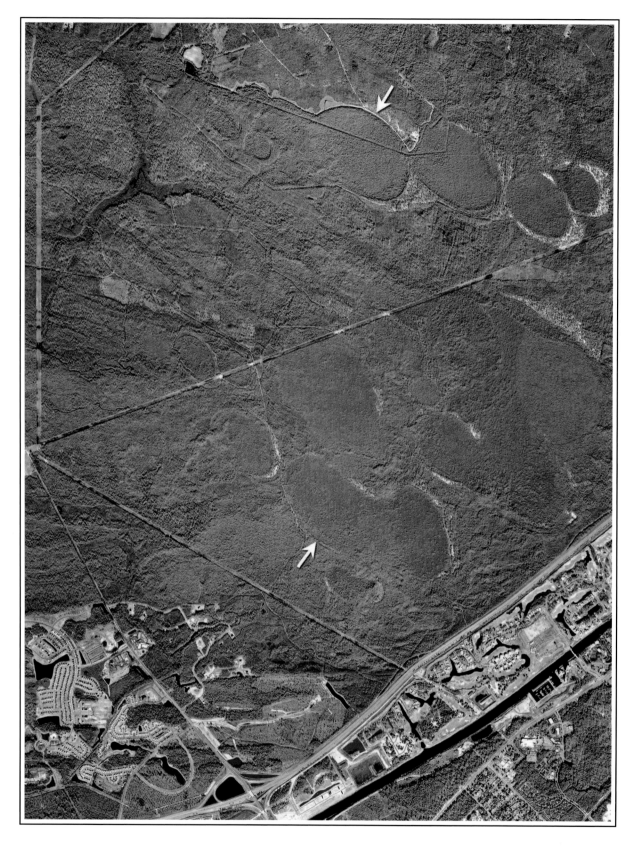

FIGURE 94. Carolina bays (arrows) in the Lewis Ocean Bay Heritage Preserve, which is located on Figure 91. Image acquired in 2006, courtesy of SCDNR.

it to form the lake. As illustrated in Figure 95A, strong winds that blew from opposite directions (southwest and northeast), presumably during different seasons of the year, generated the waves that carved the shorelines into their characteristic elliptical shapes. An example of winds forming arcuate lake shorelines is illustrated for an area on the Alaskan Peninsula in Figure 95B. Modern-day examples of features that resemble Carolina Bays and have the origin illustrated in Figure 95A have been observed by us in Tierra del Fuego and on the North Slope of Alaska. However, there is no evidence that Carolina Bays have been forming in coastal deposits (e.g., barrier island, deltas, flood plains) developed in South Carolina during the present highstand of sea level (past 4,500 years).

Many bays have sandy ridges (rims) along their southeast shoreline, which are remnant shoreline deposits composed of sand dunes and beach sediments. Recent studies by Ivester et al. (2007), who dated rim sediments of a number of "bays" in South Carolina and Georgia, found a "clustering of dates for rim formation and activity during and just prior to stadials" over the past 130,000 years. A stadial is a warmer period during a glaciation of insufficient duration or intensity to be considered an interglacial episode. Apparently, then, there were climatic fluctuations when the bays were in their formative stages during the late Pleistocene Epoch (presumably during the so-called Wisconsin glaciation and possibly even earlier), with dune formation along the bay rims being more prevalent during drier periods. Exactly what the wind conditions were during these episodes is still a matter of conjecture, but the geomorphic evidence would indicate that they were bimodal, as is indicated in Figure 95A. The wind rose given in that diagram is for the Wilmington, North Carolina airport for the last 10 years. There have been attempts to determine the wind directions during the Pleistocene, but there is no consensus on this. It would appear from the shape of the bays, which agree with wind data where modern bays are forming (e.g., Tierra del Fuego), that the winds had similar directions to the winds at the present Wilmington airport (i.e., bidirectional from the northeast and southwest), even though they may have been much stronger. Is that circular reasoning?

Another recent paper by Wright and Forman (2007), who studied two overlapping bays in Horry County (the area under discussion here), estimated that the formation of the inner rim of the bays they studied started around 20,000 years ago and that the "most recent interior bay sediments" were deposited after about 7,500 years ago. These recent studies show that the passion of scientists to learn about the Carolina Bays, which started in the late 1800s, has not died, and that we still apparently have much to learn. The same thing could be said about theories regarding the origin of the bays, though we consider that argument to be over. Nevertheless, do not be surprised if you open your newspaper some Sunday morning and see a headline such as – CAROLINA BAYS WERE CREATED AS RELIGIOUS SYMBOLS BY ALIENS FROM ANDROMEDA OVER 200 THOUSAND YEARS AGO!!

These bays in the preserve sometimes fill with rainwater, usually in winter and spring, and dry out in the summer month, a seasonal pattern that determines which plants and animals inhabit the bays. Dense thickets of plants, such as blueberries, huckleberries, fetterbush, and zenobia, cover the bays and mingle with sweet bay, catbriar, gallberry, and titi. A thin canopy of pond pine and loblolly bay trees usually surrounds the bays. In other areas, such as eastern North Carolina, the Carolina Bays contain permanent ponds and lakes, but in South Carolina, most of them are filled with thick deposits of peat (fine-grained waxy substance composed mostly of plant remains).

Prescribed burning and historic natural burns have played a major role in shaping the bay habitat and pine flatwoods, as well as preserving the natural habitats. Fire spares fire-tolerant trees, such as the longleaf and pond pine. Fire also favors herbaceous

ⒶORIGIN OF CAROLINA BAYS

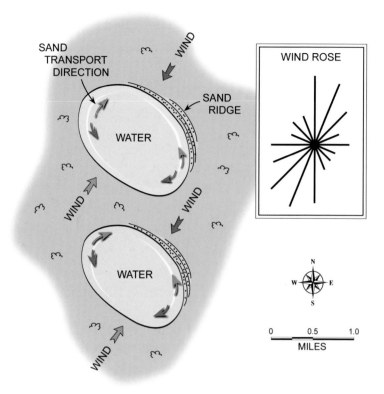

FIGURE 95. Origin of Carolina bays. (A) General model for their origin emphasizing the role of opposing wind directions. Wind rose is for the last 10 years at the Wilmington, North Carolina airport. Lines in wind rose indicate directions from which the wind blows, a distinct northeast/southwest bimodal trend. Compare this model with the real bays shown in Figure 94. (B) Modern analogs of bays like Carolina bays on the northern side of the Alaska Peninsula. In this case, only one side of the bays has developed the arcuate shoreline, because there is only one dominant wind direction (indicated by the arrows). Photograph taken in August 1976.

plants, such as fly traps, pitcher-plants, and native orchids.

Access to the bays in the preserve is difficult, because of the dense vegetation in them. However, if you become completely fascinated with Carolina Bays, better access to one is possible at **Woods Bay State Park**, which has a nice boardwalk and canoe trail within a large-sized bay. This park is located about 20 miles southwest of Florence (a short distance off I-95). If you wish to go there from Myrtle Beach and Conway, take US 378 (west). After Lake City, exit right onto SC 341 (north). Continue on 341 through Olanta. From there drive approximately 2 miles and then turn left onto Norwood Road (SC-S-43-150). At the stop sign, turn right onto Woods Bay Road (Hwy 48). The park entrance will be on the left.

Lake Waccamaw, North Carolina, a large open-water lake, is another Carolina Bay in relatively easy driving distance from Myrtle Beach. This bay also has a state park on its shore. It is located approximately 40 miles east of Wilmington, North Carolina, just off US 74.

HUNTINGTON BEACH STATE PARK

This park, which is open from 6 to 6 between October and March (evening hours extended to 10 pm the rest of the time; entry fee $5), is one any naturalist visiting the area would definitely not want to miss. It has a beautiful, well-preserved natural beach, by far the best in the Grand Strand. The park also contains 2,500 acres of natural landscape in an otherwise intensely developed shoreline. Of this acreage, 1,060 acres are salt marsh and tidal waters, 750 acres are forests, 90 acres are freshwater and brackish marshes, 400 acres are maritime shrub thicket, and 200 acres are sandy beach and dunes. With such a variety of habitats, a wide array of bird life is present, as one might expect. The free bird checklist you can obtain at the entrance booth lists 313 species, 108 of which are ranked as being either abundant or common. This checklist, which was

last revised in 2006, is based upon sightings within the park boundaries since 1966. We have birded this area several times and it is definitely one of the better places for birding along the coast.

Some of the most noteworthy birds we have seen there include, first and foremost, the **northern gannet**. We have never visited this site in the winter months that we didn't see several of them performing their vertical dives into the water from altitudes as great as 30 feet. The gannet's arrow-straight dives are in stark contrast with the clumsy, fluttering crashes and splashes into the water made by the **brown pelicans**, which almost always are part of the feeding frenzy. The best place to view the gannets is from the end of the south jetty at Murrells Inlet at the northern boundary of the park, which can be reached by walking about 1.2 miles along the beach.

Other birds of note that are usually present in the winter time include grebes, common loons, and a wide variety of shore birds (e.g., black-bellied plover, semipalmated plover, willet, sanderling, dunlin, and short-billed dowitcher). Another good time to bird this area is in the spring, when in the upland areas you might see a variety of either migrating or resident species, such as the great-crested flycatcher, eastern kingbird, yellow-rumped warbler, pine warbler, common yellowthroat, and chucks-will's widow (okay, you may only hear them). Walking the roads through the wooded areas also provides the opportunity to look for passerines.

The drive across the causeway over the marsh just past the entrance is also a good place to see a variety of birds. After you cross the causeway, take a left and you will soon come to the marsh boardwalk and education center, an excellent spot to view the marsh ecosystem. The entrance to the Sandpiper Pond Trail, which winds for 1.5 miles through a variety of habitats, is located across the road from the parking area for the boardwalk/education center.

After you visit the boardwalk, you can reach the parking area for the beach by continuing north along the road for a few hundred yards. As you walk from the parking area out to the beach, note the edge of the **wax myrtles** (*Myrica cerifera*) on the facility, which are notably salt pruned. Continuing out of the myrtles toward the beach, you cross a zone of three prograded lines of foredunes with sea oats (*Uniola*) on their crests that is about 250 feet wide. The foredune zone may be different when you visit, but it was that wide in the winter of 2007.

When we visited the park at low tide on 14 January 2007, there was a high, eroded scarp on the foredunes and the intertidal beach was very flat (see Figure 96). Large swell waves created by an offshore storm were breaking in the surf. This was different from what we had seen on this beach in years past, when a large, welded intertidal bar extended much of the way from the beach entrance to the jetty (see Figure 97). The lagoon (runnel) on the landward side of the bar was typically frequented by large numbers of shorebirds and migrating ducks.

It is beyond the scope of this book to discuss it in detail, but the two armor stone jetties at Murrells Inlet, which were built in 1977, have become the archetype for a feature called a **weir jetty**, judging by the large number of references to them in the coastal-engineering literature. As noted earlier, a weir jetty system was also installed at the Charleston Harbor jetties. A weir is a low-elevation portion of the updrift jetty (northeasterly one in this case) that allows sand to pass along the beach, over the jetty and into a catchment area, usually a deepened hole located just inside the jetty. The sand that accumulates in the catchment area can be periodically dredged out and passed to the downdrift (southwest) side of the jetty. As noted earlier, according to Tim Kana the U.S. Army Corps of Engineers does this periodically at this inlet using a hydraulic dredge. We don't have all the details on how frequently this has been done, but the wide, generally prograding beach on the downdrift side of the jetties (inside the park) would seem to be evidence that the system is working. The aerial photograph in Figure 97 shows a large

accumulation of sand on the southwest (downdrift) side of the jetties on 10 July 1989, about 12 years after the jetties were built. Furthermore, Litchfield Beach, to the south of the park, has been one of the most stable beaches in South Carolina for the past 50 years (Tim Kana, pers. comm.).

Another aspect of the engineering design of these jetties was the construction of a fisherman-friendly walkway along the top of the south jetty, out to near its end. Thus, fishing has become an added attraction of this great state park. Judging by the large number of fishermen we normally see there, the fishing must be pretty good. Surf fishing is also a popular pastime within the park.

BROOKGREEN GARDENS

The entrance to Brookgreen Gardens, a National Historic Landmark, is located immediately across US 17 from the entrance to Huntington Beach State Park. We have never spent any significant time in this facility, but it is no doubt a unique setting that would be worth the visit should you be so inclined.

As noted on the Gardens' web site, "the thousands of acres (9,000 to be exact) in Brookgreen's Lowcountry History and Wildlife Preserve are rich with evidence of the native plants and animals of the South Carolina Lowcountry, as well as the great rice plantations of the 1800's."

Some of the many additional attractions of the Gardens include:

- A boat excursion
- Nature trails
- A zoo, advertised as the best on the coast of the Carolinas
- Significant collections of figurative sculpture in an outdoor setting (one of the original owners, Anna Hyatt Huntington, was a noted sculptress)

Brookgreen Gardens is open daily from 9:30 am to 5 pm, with a few exceptions. The entry fee is $12 for adults.

FIGURE 96. Beach at Huntington Beach State Park. View looks southwest. Photograph taken at low tide on 14 January 2007.

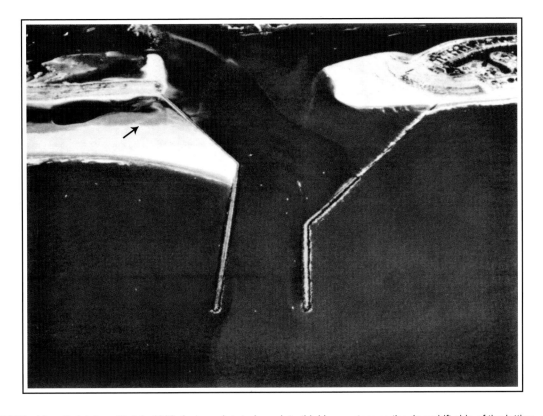

FIGURE 97. Murrells Inlet on 10 July 1989. Arrow points to large intertidal bar system on the downdrift side of the jetties.

PAWLEYS ISLAND

To drive to Pawleys Island, a small formerly prograding barrier island with active small tidal inlets on both ends, go south from the entrance to Huntington Beach State Park on US 17 for about 7 miles. This island, which is illustrated by the vertical image in Figure 98, has been settled for decades and has been a favorite beach vacation spot for local South Carolinians over the years. The island is famous in the folklore about the state. For example, a local legend on the island has grown about *The Gray Man*, a friendly ghost who warns of impending hurricanes (Prevost and Wilder, 1972; Vaughan, 1975).

Pawleys Island is not a gated community, thus access is not a problem. There are two causeways to the island that cross the relatively narrow marsh in the backbarrier area. A parking area on the south end of the island provides access to the southern inlet (Pawleys Inlet).

The following succinct summary of the island's beach management history is given on the following web site – www.csc.noaa.gov/products/sccoasts/html/pawl.htm#beach.

Pawleys Island is located in Georgetown County between Pawleys and Midway Inlets. The island has 3.5 miles of beachfront and is .5 mile wide. Groin fields were constructed in the late 1940s to counteract erosional trends and in response to various storms over the years. The southern portion of Pawleys is low-lying, with little or no sand dunes. The central portion of the island has some of the highest dunes in the state, while the northern, accretional end has a wide field of low dunes.

Hurricane Hugo took its toll on the beach and development on Pawleys Island. Areas landward of large dunes were protected from overwash. Low dunes also survived, probably because they were rapidly flooded and overtopped by the storm surge. More than 3 miles of shoreline required

emergency beach nourishment following the storm. Development at the south end was severely impacted and 27 of the 29 houses located there were destroyed. A channel breached the southern spit but was consequently filled and the houses rebuilt.

As noted in this account from the NOAA web site, the formerly prograding central part of the island is high and was not washed over during hurricane *Hugo*, but the dunes showed significant erosion. As also noted, there has been some beach nourishment on the island. An "emergency" amount of 220,000 cubic yards was added in 1990. This is close to the amount that normally passes along the coast in the longshore transport system in one year. In 1998, an apparently larger amount of sand was borrowed from the bare sand spit that projected beyond the developed zone on the southern end of the island. This sand was placed further north along the island. At the same time, the groins that had been present on the outer beach for decades were rehabilitated.

As discussed in some detail in the section on hurricane *Hugo*, the low-lying, washover terrace of the spit at the south end of the island was severely washed over during the storm (see the sketch map we made of the island soon after *Hugo* in Figure 64A). At the present time (our last visit was in early 2007), this low spit has been completely rebuilt with closely spaced, presumably fairly expensive houses (Figure 65).

There is a parking area at the south end of the developed zone where you can take a walk along the beach to Pawleys Inlet. This is a pleasant spot and a good place to look for shorebirds around the inlet entrance (at low tide).

Another beautiful natural area is the downdrift offset inlet at the north end of the island (Midway Inlet). To get there, drive along the back island road for 0.3 miles north of where the north causeway extends across the marsh. At that point, there is a parking area from which you can walk several

FIGURE 98. Vertical infrared aerial photographic mosaic of Pawleys Island acquired in 2006. Courtesy of SCDNR.

hundred yards along the beach to the inlet. The two inlets on either side of Pawleys Island are shown by the vertical satellite image in Figure 98.

12 THE SANTEE/PEE DEE DELTA REGION

INTRODUCTION

The somewhat arbitrary boundaries we have chosen for this compartment are the southern end of Pawleys Island to the north and the southern end of Bulls Bay to the south (see Figure 99). This area bounds the whole of the shoreline that is affected in some way by the two major piedmont rivers, the Pee Dee and the Santee, that have combined over time to create the largest deltaic bulge on the east coast of the United States.

For purposes of discussion, we have further subdivided this compartment into four subcompartments:

1) **The Winyah Bay Estuary Complex** – This subcompartment includes at its northern boundary the upscale gated community of DeBordieu, which adjoins the North Inlet-Winyah Bay National Estuarine Research Reserve at its southern border. The Reserve contains the pristine tidal inlet/estuarine complex of North Inlet, as well as part of the shoreline of Winyah Bay. The entrance to Winyah Bay itself, which receives freshwater influx from four rivers, is stabilized by two long jetties.

2) **Santee Delta** – The bulge in the shoreline in this part of the delta was created by the Santee River, which carried to the coast one of the heaviest sediment loads of any river on the east coast before the construction of a dam that created the Santee Cooper Lakes in 1942. Much of the delta's surface is presently preserved as natural areas. The large Tom Yawkey Wildlife Center Heritage Preserve (31 square miles) occupies the most shoreward land area between Winyah Bay and the North Santee River outlet. The Santee Coastal Wildlife Management Area (WMA), about 2,000 acres of marsh and swamp, is located between the North and South Santee River channels, and the Santee Coastal WMA occupies an area of almost twice that to the south of the mouth of the South Santee River.

3) **Cape Romain** – This major cuspate foreland, which is about 12.5 miles in length along the outer shore, is the southernmost of a series of four of these triangle-shaped headlands that project out into the ocean between the central North Carolina coast and Charleston (Cape Hatteras, Cape Lookout, Cape Fear, and Cape Romain, from northeast to southwest). The outer beach of the foreland, as well as the marsh back to the Intracoastal Waterway, are located within the Cape Romain National Wildlife Refuge.

4) **Bulls Bay** – This area is an open, arcuate-shaped bay with an entrance about 7 miles across that

FIGURE 99. Vertical infrared image of the Santee/Pee Dee Delta compartment acquired on 11 May 2000, courtesy of Earth Science Data Interface at the Global Land Cover Facility.

contains no barrier island or sand beaches. The landward side of the bay contains an unbroken band of salt marsh.

THE WINYAH BAY ESTUARY COMPLEX

Description

A review of the rather extensive literature on this subcompartment reveals the following major characteristics of Winyah Bay itself:

- Three major rivers, the Pee Dee, Waccamaw, and Black, empty into this estuary.
- The watershed for these rivers is approximately 18,000 square miles, one of the largest watersheds on the east coast south of the Chesapeake Bay (see Figure 100).
- More than 16,000 square miles of this drainage area are associated with the Pee Dee/Yadkin River system, which originates in the Blue Ridge Mountains area of North Carolina.
- There are some conflicting numbers in the

156

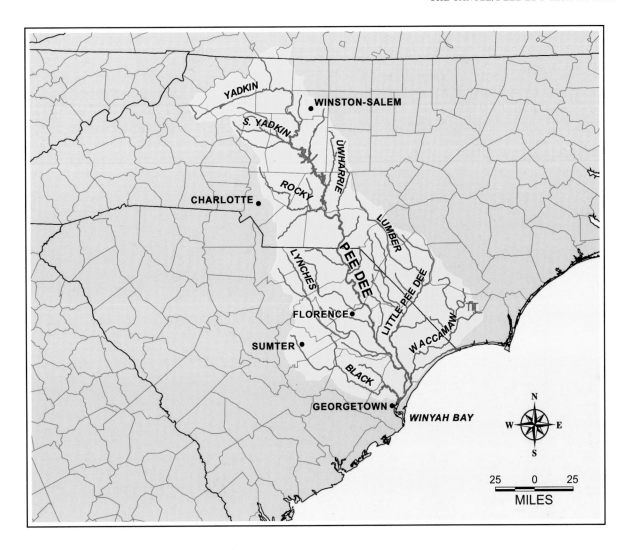

FIGURE 100. Watersheds of the rivers that empty into Winyah Bay.

literature, but the consensus appears to be that under high, freshwater flow conditions, this estuary, the fourth largest estuary on the east coast of the U.S., has a classic salt-wedge structure, but during lower flows, it is partially mixed. Some authors say it is vertically homogeneous under extremely low freshwater input and higher than normal tides.

- It has a complex bathymetry, with bifurcated channels and wide tidal flats (Figure 99).

Places to Visit

If you are driving south along US 17 from Pawleys Island, at 7 miles you arrive at the visitor center for the Hobcaw Barony, located on the left. This is also the entrance to the Baruch Marine Field Laboratory of the Baruch Institute for Marine and Coastal Sciences that was established in 1969 through the joint efforts of the Belle W. Baruch Foundation and the University of South Carolina. The Field Laboratory, which some of Hayes' students used while he was a Professor at the University, provides facilities for research and education on estuarine, freshwater, and nearshore ocean systems. It is possible for the general public to participate in a variety of educational programs and tours on the facility that deal with such topics as the birds and the fish in the area. Because enrollment is limited for most activities, registration by phone is required.

The number to call is (843) 546-6219 extension 0. More information is available on the web site for the Baruch Institute (http://links.baruch.sc.edu/Data/).

This site contains a small, but very pristine, estuarine complex that connects to the ocean through North Inlet, a natural tidal inlet that has never been "improved" by man. North Inlet is part of the Long-Term Estuarine Research (LTER) Program of the National Science Foundation. Note the large ebb-tidal delta of the inlet in the vertical image in Figure 101.

This intriguing site is definitely a great place for a naturalist to visit, if your and the Institute's schedules allow it.

As you leave the Barony visitor center, go west for about a mile and a half on US 17, where you cross the bridge over the Waccamaw River portion of the upper Winyah Bay estuary. Before you drive over the next bridge, which crosses the Pee Dee/Black Rivers portion of the upper Winyah Bay estuary, turn right into the **Hobcaw Point Observation Site and Fishing Pier**. From the parking area, you can walk out on a remnant of an old bridge that is now a fishing pier. As you walk out on the bridge, you will cross a fringe of *Phragmites* and *Spartina cynusuroides,* a spectacular example of a low salinity brackish marsh (with zonation). This is also a good birding spot, and, by the way, it is about the only place you can observe the Winyah Bay estuary without the services of a watercraft.

Historic downtown Georgetown, which you

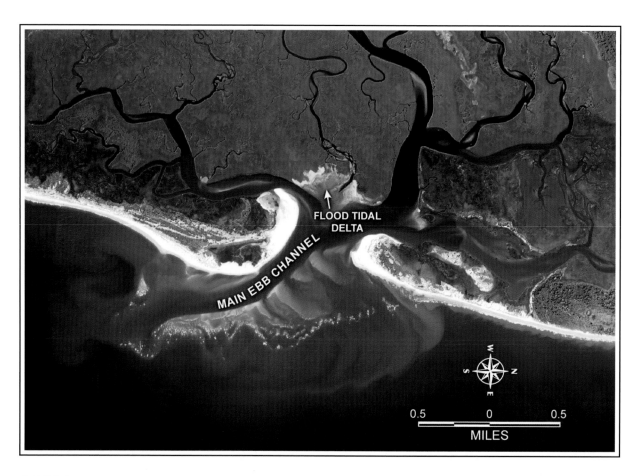

FIGURE 101. Vertical infrared image of the North Inlet area acquired in 2006, courtesy of SCDNR. Note the huge ebb-tidal delta and small flood-tidal delta, a characteristic of mesotidal coasts like South Carolina. Like several other tidal inlets in South Carolina, the main ebb channel of the ebb-tidal delta has been oriented to the south as a result of the dominant northeasterly waves.

will be close to after you cross the second bridge (heading south), is a nice place to visit. If it is open, the Rice Museum, located in the historic waterfront district, should not be missed.

This town, as well as historic Charlestown, owe their early wealth and prominence to a form of agriculture unique to this area of North America in the 17th–19th centuries, **rice farming**. Rice growers along the South Carolina coast began exporting rice in the 1690s, with Winyah Bay and the Santee Delta regions being two of the pioneering areas. By 1850, growers in South Carolina were exporting 160 million pounds of rice per year. This would not have been possible without the enormous amount of slave labor involved, with the African slaves also being credited with importing their rice-growing knowledge and experience from their native lands.

In the earlier years, the rice was grown in upland swamps, but over time, the nutrients in the soils of the swamps were rather quickly exhausted by the rice plants. By the mid 1700s, the technique was switched to **tidal rice fields**. In this method, ponds were built with gates that could control tidal waters flowing into and out of the rice-growing ponds, with the ponds being flooded with fresh water during the growing season and the water being released from the ponds on falling tides to make harvesting the crop possible. This was an ingenious method that used the changing tides for irrigating and draining the rice fields to maximize production and reduce labor. Furthermore, with the river continuously replenishing the pond soils with nutrients, the soil never became exhausted. Therefore, the requirements for the use of this method included two major ingredients: 1) incoming fresh water, and 2) a considerable tidal range to allow the necessary water exchange. The limiting factors were the diminishing tidal range as one approached the "head of tides," which determined the landward limit of the rice ponds, and the encroachment of salty sea water up the river, which determined the seaward limit. Many of the estuarine systems along the South Carolina

coast met these requirements perfectly, with the rice culture zone extending 15 miles, or more, up the rivers in places. Observe a topographic map of the marsh systems of any major South Carolina estuary today and you will see the extensive, square relict ponds along all of their upper reaches.

Due to the great success of rice cultivation along the coast, especially the tidal rice fields, the relatively small number of rice growing families in South Carolina soon became extremely wealthy, as evidenced by the mansions they built on the plantations, and the even greater mansions built by the growers and their service groups (e.g., merchants) in downtown Charleston. With the loss of slave labor resulting from the Second Civil War (aka The War Between The States), the rice culture business went into a steep decline, with its final demise in the Santee Delta and Winyah Bay areas being brought about by hurricanes in the early 20th century.

THE SANTEE DELTA

General Morphology and Sedimentation Patterns

This discussion will focus mostly on the Santee River component of the Santee/Pee Dee delta complex. Late Pleistocene coastal terraces made up of multiple, parallel highstand beach ridges lie adjacent to the main body of the delta (see Figure 99). These upland Pleistocene sediments were dissected by the Santee River during the Late Wisconsin sea level lowstand, resulting in the incision of the river valley. Whether or not this valley was occupied by the river during any of the earlier glaciations has not been determined as yet. However, we do know that basal river sands grading upward into deltaic sand and muds progressively filled the drowned river valley as it was flooded during the last major sea-level rise, eventually creating this limb of the largest river delta on the east coast of the United States.

As shown on the satellite image in Figure 102, for purposes of discussion, we have subdivided the delta into three components – upper delta plain, lower delta plain, and delta front. The **upper delta plain**, the landward limit of which coincides roughly with the "head of the tides," contains the highly sinuous (meandering) main channel of the Santee River as well as a number of abandoned channels. Point bars composed of coarse-grained sand are present along the main channel, as well as abandoned oxbow lakes with minor connections to

it, such as the one shown in Figure 103. Extensive cypress/tupelo swamps and bottomland hardwoods occur throughout the upper delta plain, some of which contain peat deposits. Many of the coal deposits in Kentucky and West Virginia were formed in similar environments. Back in the late 1970s, geologists who went on our coastal field trips, which always included a day on the Santee delta, followed up our trip with one to the Kentucky/West Virginia coal fields, led by our friend and former partner, Dr. John C. Horne.

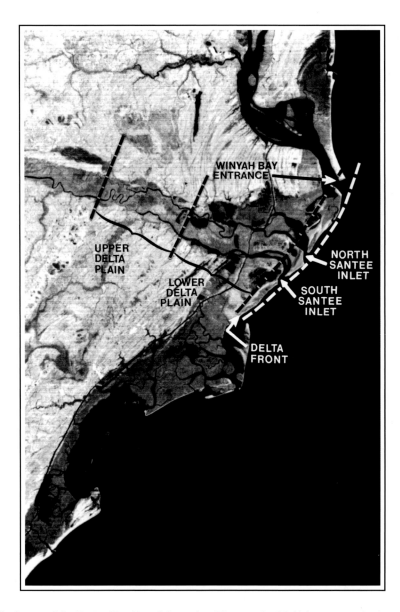

FIGURE 102. Satellite image of the Santee/Pee Dee delta region. Three major tidal inlets are marked, as well as the designated delta components.

FIGURE 103. Abandoned oxbow lake in the upper delta plain of the Santee portion of the Santee/Pee Dee Delta located just off the main channel of the North Santee River. Since this oxbow was cut off, probably at least a few hundred years ago, it has filled with over 30 feet of fine-grained sediment (clay and silt). Note the freshwater marsh plants (arrow; mostly saw grass, *Cladium jamaicenci*) that have grown on the surface of the muddy sediment. Infrared photograph taken in April 1977.

About 15 miles inland from the delta front, and within the upper delta plain, the Santee River bifurcates into two systems, the South Santee River and the North Santee River (Figures 99 and 102) . Historically, the major volume of flow has switched from one system to the other, but at the present time, the bulk of the fresh water flows down the North Santee (73% according to Nixon, 2006).

The **lower delta plain** is a 6-mile-wide triangular belt of mud-dominant habitats. This muddy zone separates the more sand-rich upper delta plain from the sand-dominant delta front. Before rice cultivation started on the delta (in the late 1600s), a freshwater swamp extended to near the present North Santee River Inlet. Today, however, the surface of the lower delta plain is a checkerboard pattern of refurbished rice ponds encased in brackish and salt water marshes (see Figure 104). Three major channels, called North Santee Bay, North Santee River, and South Santee River, cut across the lower delta plain of the Santee region (Figures 99 and 102). Salt marsh occurs much further upstream in the South Santee River than the North Santee, because it receives so little fresh water influx from the main stem of the Santee where it enters the delta. Coarse-grained river sand appears to be moving through the lower delta plain along the North Santee River channel bottom to the delta front; however, this part of the delta is dominated by **mud deposition**. This mud deposition presumably is caused by the presence of the *zone of the turbidity maximum* in the region where the river water mixes with the salt water brought in by the flood tides. No detailed suspended sediment studies have been

FIGURE 104. Former rice ponds in the lower delta plain of the Santee portion of the Santee/Pee Dee Delta that are presently maintained for waterfowl by the U.S. Fish and Wildlife Service. Photograph taken on 5 May 1997.

carried out in the Santee delta area. However, a study by Zabawa (1976) indicates that "trapping of fine-grained sediment by the estuarine circulation pattern" does occur where the Pee Dee and its associated rivers enter Winyah Bay.

As shown by the vertical images in Figures 99 and 102, and the oblique aerial photos in Figure 105, marine processes have molded the **delta front** of the Santee/Pee Dee delta into a huge, arcuate sand-dominant bulge in the shoreline. Three major tidal inlets – the Winyah Bay entrance, North Santee River Inlet, and the South Santee River Inlet – are separated by recurved spits and by preserved, older beach ridges. Large ebb-tidal deltas and small flood-tidal deltas occur at both of the Santee Inlets. The Winyah Bay entrance, which has no flood-tidal delta, is modified by jetties built around the turn of the century.

Offshore of the delta lies a complex of reworked deltaic deposits. The transgressed subaqueous delta extends 12 miles seaward of the active delta front, exhibiting in places relief in excess of 40 feet over a distance of 0.3 miles (see Figure 90A).

Impact of Man

Since the beginning of historic times, the upper surface and river flow of the Santee arm of the delta have been modified dramatically by man. As noted earlier, from the late 1600s to the late 1800s, extensive rice farming was carried out on the delta, practices that modified the natural sediment and vegetation patterns. The primary impact was the deforestation of the large cypress/tupelo swamp that covered much of the surface of the lower delta plain. The following excerpt from the journal of

FIGURE 105. Oblique aerial views of the delta front of the Santee/Pee Dee Delta. (A) Infrared photograph by P.J. Reinhart taken between mid and high tide in October 1979. Note the development of beach ridges along the seaward margin of this mixed-energy delta (arrows). The channel in the middle ground is the North Santee River. Note the slightly exposed swash bar at the river mouth. Winyah Bay can be seen in the far distance, and the south Santee channel is visible in the lower left. (B) Photograph of the mouth of the North Santee River at low tide on 5 May 1997, eleven years after the rediversion of approximately 50% of the river flow back into the main channel of the river. The huge swash bar in the middle distance, which eventually welded onto the beach, probably derived much of its sand from the revitalized river.

John Lawson (Lawson, 1709) describes the mouth of the North Santee River in 1700:

> *"As we row'd up the River, we found the land towards the mouth, and for about sixteen miles up it, scarce any thing but swamp and percoason, affording vast ciprus-trees, of which the French make canoes, that will carry fifty or sixty barrels, …. For although their river fetches its first rise from the mountains, and continues a current some hundreds of miles ere it disgorges itself, having no sound bay or sand banks betwixt the mouth thereof, and the ocean. Notwithstanding all this, with the vast stream it affords all seasons, and the repeated freshes it so often alarms the inhabitants with, by laying under water great part of their country, yet the mouth is barr'd, affording not above four or five foot water at the entrance."*

These great "ciprus trees" were cut down later in the 1700s, and ponds were dug for rice cultivation. Presently, many of these ponds are maintained by the U.S. Fish and Wildlife Service and private landowners for wintering waterfowl habitat (Figure 104). The "barr'd" entrance to the river mouth is a **large ebb-tidal delta**, a feature commonly found off the inlets of the mesotidal South Carolina coast, but one completely missing from the macrotidal coast of Great Britain, Lawson's native home.

Construction of hydroelectric dams further upriver in the early 1900s steadily reduced the sediment supply to the delta, culminating in 1942, when a large dam was created that diverted 90 percent of the original discharge of the Santee River into the Cooper River. The historical mean annual flow of 18,500 cubic feet per second (cfs) was reduced to 2,200 cfs (Kjerfve and Magil, 1989; 1990). After the diversion, most of the delta front shoreline started eroding rapidly, primarily as a result of the decreased supply of sediment to the delta. In 1986, a variable percentage of the flow of the Santee River above the dam at Lake Marion (only

rarely exceeding 50 percent of original discharge) was rediverted back to the river from Lake Moultrie via a newly constructed rediversion canal in order to, at least partially, solve sedimentation problems in Charleston Harbor resulting from the original diversion of the river.

These two major man-made modifications to the flow of the Santee River have had a marked effect on the salinity of the water in the delta. The plant composition of the marshes that populate the surface of the lower delta plain also underwent a major change, as well as did parts of the upper delta plain, though swamps are more common in the upper reaches of the delta. Zach Nixon, a long-term employee of RPI who completed an M.S. thesis at Duke University on the delta, modeled surface water salinity of the delta for three key periods: 1) pre-dam (1875-1941); 2) post-dam (1941-1986); and 3) post-rediversion (1986-2003). As expected, with 90% of the fresh water that formerly came into the delta region diverted elsewhere, there was a massive intrusion of salt water into the lower delta plain region after the diversion in 1942, resulting in the encroachment of salt water plants up the river. After the rediversion, the freshwater conditions returned to the delta, but not quite to the level they were at before the dam was built.

With regard to the plant communities, Nixon concluded: *"Overall, there is strong evidence for fairly dramatic change in the spatial distribution of emergent marsh communities of the Santee River Delta between 1941 and 2004. This change has been in the form of an overall shift from freshwater-adapted species and communities to more saline species and communities, though this is not true across the entire Delta. The greatest magnitude of change is located in a band running parallel to the outer coast approximately 3-4 kilometers northwest of the Atlantic Intracoastal Waterway."* According to Nixon (2006), whose thesis also involved surveying the marshes in the delta, the following pattern of dominant marsh species now exists:

1) Near the delta front – extensive smooth cordgrass (*Spartina alterniflora*) and black needlerush (*Juncus roemerainus*) monotypic marshes dominate.

2) Further up the lower delta plain – giant cordgrass (*Spartina cynosuroides*) dominates, often in large stands.

3) Further upriver near and possibly into the upper delta plain (especially on the South Santee arm) – more diverse communities dominated by North American cattails (*Typha* spp.), *Schoenoplectus* spp., watergrass (*Echinochloa* spp.), spike rushes (*Eleocharis* spp.), wild rice (*Zizaniopsis* spp.), and panic grass (*Panicum* spp.) occur.

4) Even further upriver, near the upper reaches of the delta, the dominant marsh grass becomes sawgrass (*Cladium jamaicense*).

Places to Visit

First we will describe a proposed driving tour of the large headland that lies between the North Santee River and Winyah Bay, the path of which is given on Figure 106. Drive south of Georgetown on US 17 and cross the bridge over the Sampit River. After crossing the bridge, go about a mile and turn left on South Island Road. Once on that road, it is about 8 miles to the South Island Ferry, an important destination for a naturalist. The first 3 miles or so along the road are through a developed area, but eventually you turn to the southeast and cross a series of prograded Pleistocene beach ridges of the Myrtle Beach Pleistocene formation, which is said to be roughly 200,000 years old. The old beach ridges are separated by swales that contain swamp vegetation. These closely spaced beach ridges, which show up clearly on the image in Figure 106, prograded seaward for at least 7 miles during one or two of the interglacials, but, at this point, we are uncertain about the exact timing of these events. The morphology of this broad, arcuate projection of beach ridges is very similar to that of a

wave-dominated delta, the general model of which is given in Figure 42B. To us, this means that the average wave height along this coast was much greater when that highstand delta was forming than it is today.

Continue along the drive until you reach the South Island Ferry. This ferry crosses the Atlantic Intracoastal Waterway, which, at this location, was dredged through the prograded Pleistocene beach ridges of the Myrtle Beach formation. The ferry provides access to the **Tom Yawkey Wildlife Center Heritage Preserve**, considered to be one of the most outstanding gifts to wildlife conservation in North America. This wildlife center and preserve was willed to SCDNR in 1976 by the late Tom Yawkey, former owner of the Boston Red Sox. The center's web site notes that the preserve is composed of 31 square miles of marsh, managed wetlands, forest openings, ocean beach, longleaf pine forest, and maritime forest. It is principally dedicated as a wildlife preserve, research area, and waterfowl refuge.

Visitation to this fabulous site is by guided tour only. The preserve biologists run two field trips a week (a guided driving tour) for 14 people per trip. To sign up call 843-546-6814, but you have to do it MONTHS IN ADVANCE! We called in January 2007, just to check out the procedure, and they told us they were full up until June 2007 (just in case you were wondering). Hayes and his students carried out research in this area about 25 years ago. It's great, definitely worth the effort.

To continue the drive, go back down South Island Ferry Road for a few hundred yards and turn left on Estherville Road. Heading south, the road now runs through the Pleistocene beach ridges, which are covered with pines for about 2.5 miles. The road takes a right-angle turn to the right at the entrance to Annadale Plantation. Continuing west, you soon come to Millbrook Plantation, which has an impressive exterior. These plantation houses like this one are not nearly as elegant as the houses in downtown Charleston, with small claustrophobic

FIGURE 106. Vertical infrared image of delta front portion of the Santee/Pee Dee Delta acquired on 11 May 2000, courtesy of Earth Science Data Interface at the Global Land Cover Facility. The path of a recommended driving tour through the area of the multiple Pleistocene beach ridges is shown.

rooms and small windows. However, they usually have beautiful exteriors and the grounds are always fantastic. Continuing for another two miles or so you pass the entrance to Kinloch Plantation on the left, which has elegant grounds. This plantation was formerly owned by media mogel Ted Turner. In the 3 miles left between Kinloch and US 17, you will pass by the entrances to two more plantations, Woodside and Rice Hope. Most of these plantations along this side of the North Santee River maintain

the old rice ponds for waterfowl.

If you take a left on US 17 you will reach the North Santee River in a little over half a mile. On the left is the Pole Yard Boat Ramp, the place where we offloaded the students at the end of the field day during the professional field seminars that we conducted for the American Association of Petroleum Geologists, Schlumberger Well Services, and other clients from the oil-and-gas industry for 30 years (1976-2005). During those field seminars, we always came to the Santee Delta, which means we have ridden up and down this river in our catamaran more than 150 times over the years. During that period, over 3,000 geologists from all over the world, most notably from western Canada, listened (hopefully) as we told the story of the geology of the delta, which is partly outlined above. But that was only part of the story. Thinking back, we have many fond memories of those boat rides, some of which follow:

- The birding on this delta is in a class by itself for this part of the world. Some of the most notable birds in the upper reaches of the delta are the bald eagle, Mississippi kite, swallow-tailed kite, numerous nesting ospreys, and wood ducks. The lower delta contains myriads of marsh and shore birds, including royal and Caspian terns, laughing gulls, and many others. Christmas bird counts carried out on the delta have some of the highest numbers of species of any location east of the Mississippi River. We went on one of the Christmas counts and were amazed at the numbers of waterfowl, measuring in the thousands, that we saw on the delta that winter.

- Wading in the swamp in the upper delta plain and telling stories about the **Swamp Fox** (Francis Marion, the original guerrilla fighter in the First Civil War; aka American Revolution), repeating how Lord Banastre Tarleton gave him his nickname. Prowling the rim of the black muck in the swamp, Tarleton was heard to say – "Come boys, let's go back and fight the Gamecock (General Sumter). As far as this damned old fox is concerned, the devil himself couldn't catch him."

- A sturgeon 3 feet long jumping into the boat.

- Thunderstorms! Abandoning the boat to gather in a fisherman's shack to avoid the lightning bolts (more than once).

- Going far up the river into the "mysterious green world" of the abandoned oxbow lakes.

- Discovering a previously unreported historic relic, an ocean-going cargo vessel built in the mid 1700s. This boat was probably beached in the marsh during a hurricane and eventually covered with mud. We found it after it was exhumed on the beachfront during an erosional episode.

- Pointing out the imprint of hurricane *Hugo* on this totally undeveloped shoreline.

Guess we don't have to tell you that we think this would be a great place to launch watercraft for a ride up into the magic green world of the upper delta plain of the Santee Delta. If you do, however, keep an eye on the tides. If you like to fish for bass and/or catfish, take a rod along.

Just across the road from the boat ramp is the **Hopsewee Plantation**, a South Carolina National Historic Landmark, which is the birthplace of Thomas Lynch Jr., a signer of the Declaration of Independence. Visiting hours are 10–4, Monday through Friday and also by reservation.

The house, which is still a private residence, is described on its web page as *"a typical low country rice plantation dwelling of the early eighteenth century with four rooms opening into a wide center hall on each floor, a full brick cellar and attic rooms. The house has a lovely staircase and there is hand carved molding in each room and random width heart pine floors are almost one and one half inches thick."* And so on. In fact, as noted earlier, you can see much more elegant architecture in downtown Charleston, but this plantation's history and grounds, with a couple of trails, are worth the visit. The house is perched on a relatively high Pleistocene terrace a short distance from the channel of the North Santee River. Amazingly, it has survived the numerous hurricanes that have struck this area over the past 300 years.

Continuing west from Hopsewee on US 17, you have about a 2.5 mile drive across the delta, essentially along where we place the boundary between the lower and upper delta plain. Just after you cross the bridge over the North Santee channel, you can pull off on the left and visit a trail across a portion of this part of the delta. The South Santee channel hugs the bank along the southwest side of the delta, and extensive wetlands, some of which are in the old rice ponds, occur between the two main river channels.

The next place we recommend you visit is **Hampton Plantation State Historic Site**. Continuing south on US 17 after you cross the South Santee River for about a mile and a half, turn right for a two-mile drive through pine forests to the Plantation grounds. Mansion hours are 12-4, Thursday through Sunday. If you are into architectural design, this house, which is shown in Figure 107A, won't thrill you much, but the grounds are exquisite.

A historic note on one of the signs on the grounds follows:

"Protestant immigrants from France, called Hugenots, were among the earliest settlers in the 1670 Carolina colony at Charles Town. Many of these early French settlers moved into the frontiers of the Santee Delta, where they retained their cultural heritage and language for many years. For this reason, the Santee Delta region became known as the French Santee. The house at Hampton was the largest great house in the French Santee region."

This Georgian style mansion, which was built sometime between the 1730s and the 1750s, is another product of the wealth created by the tidal rice farming. The plantation was owned by several prominent South Carolina families in the past, and it was once visited by President George Washington, who is said to have advised the lady of the house not to cut down a small live oak (*Quercus virginiana*) in front of the house, because they "grow so slow." That famous tree now partially blocks the view of the mansion (Figure 107A).

The last owner of Hampton was Archibald

Rutledge, the *poet laureate* of South Carolina between 1934-1973, who immortalized the site and its inhabitants in his numerous writings about life in the delta region. He was awarded the John Burroughs Medal for "the best nature writing in America." Two of his more noteworthy books are *Home By The River* and *Santee Paradise; The Beautiful Wilderness Around Hampton Plantation*.

It is a pleasant experience to walk around the garden designed by Archibald Rutledge himself, who wrote – *Like the children of the wild, I am truly nature's child.* Come in winter to see the wide array of camellias and in the spring for the azaleas. Spring would be the best season for birding, when you could listen to the songs of the nesting warblers – parula, pine, prothonatary, etc. You also wouldn't want to miss the melodic calls of the **wood thrushes** at dusk, our favorite.

Archibald Rutledge wrote a poem about the wood thrush, published in his book *Deep River*. The first two verses of that poem follow:

About the setting of the sun,
When fevers of the day are done,
A wood thrush in the tawny light
Is caroling of coming night
Upon the harp he harps upon.
The cares that to the day belong
He sings away with music song.
His notes are few, but pure and deep.
Like magic music heard in sleep,
Like love that sorrow has made strong.

Archibald was laid to rest on 15 September 1973, near the plantation house at a grave site on a bluff surrounded by large live oaks. His tombstone reads "Here he is laid to rest at Hampton his beloved home by the river." One of his most famous quotes is – *One of the sanest, surest, and most generous joys of life comes from being happy over the good fortune of others.* Come to think of it, in this case, our good fortune would be to hear the warblers singing in Archibald's garden on a spring morning and the

FIGURE 107. Plantation houses. (A) Hampton House, in the Santee/Pee Dee delta region. (B) Grove Plantation, in the ACE Basin. Both photographs were taken in the winter of 2006/2007.

wood thrush singing at dusk.

Naturally, Archibald wrote a poem entitled *The Garden I Made*, which is also published in his book, *Deep River*. A portion of that poem:

> *Whenever I think of*
> *Life's lights that must fade,*
> *The gleam and the glow*
> *That must pass into shade;*
> *When hopes that I had*
> *Now but leave me afraid—*
> *With joy I remember*
> *The garden I made.*
>
> ....
>
> *The garden I made*
> *Will keep blossoming on*
> *When life with its fevers*
> *Is faded and gone.*

Growing up in the mountains of North Carolina in the 40s and early 50s, Hayes and his pal Adrian used to read Archibald's stories in the outdoor magazines of the time. Stories about giant alligators and rattlesnakes. Stories they loved but thought were a big bunch of lies, that is, until Hayes spent some time on the Santee Delta himself.

Once you leave the plantation, about a mile and a half on up the road to the north is a canoe/kayak access point on Wambaw Creek, a great paddle all the way out to the upper reaches of the South Santee channel.

We recommend one last spot to visit in the Santee Delta subcompartment, the **Santee Coastal Reserve.** To get there, go back to US 17 cross over it and follow the signs to the Reserve. Along the way, you will pass by the entrances to the Wedge and Harrietta plantations, but they are not open to visitors. After driving into the Reserve, you pass through a savannah type pine forest, with multiple blue bird houses along the way. The main road is a drive along a Pleistocene highstand barrier complex, which contains some red-cockaded woodpecker trees. At 1.9 miles, there are some large live oaks,

scattered at random, and at 2.6 miles is a complex of buildings where the office is located.

While driving along the trails in the Reserve, mostly you pass through more pine forests. There are some ponds where we saw a big flock of blue-winged teals the last time we visited the preserve. It is probably worth a visit for the drive and the birds, depending upon the season.

CAPE ROMAIN

Morphology and Origin

Cape Romain is southernmost of four **cuspate forelands** on the mid-Atlantic coast – Capes Hatteras, Lookout, Fear, and Romain (see Figures 108 and 109). These forelands are triangular-shaped features composed of sand and/or gravel that project either perpendicular or at a small angle to the overall trend of the shoreline. If you travel the globe observing shorelines as we have done, you will find that cuspate forelands are fairly ubiquitous features. For example, two of the most striking ones we have seen occur on both sides of the east entrance to the Strait of Magellan. Perhaps the most famous one is the Dungeness headland on the south shore of Britain, a huge triangle of land composed mostly of pebbles that projects out into the English Channel. It is also a great birding spot.

Cuspate forelands are most common on shorelines that are parallel to two opposing wind directions. Similar to Carolina Bays, the two opposing wind directions generate waves that play a role in molding the sediments into the characteristic morphology, in this case triangular forelands, which typically occur at equally spaced intervals along the shore. As proposed by the pioneering Russian coastal geomorphologist, V.P. Zenkovitch (1967), waves striking a shoreline at a large angle are capable of high rates of longshore sediment transport. As the sediment moves along shore, the longshore transport system becomes "saturated" (Zenkovitch's term) and features such as rhythmic

170

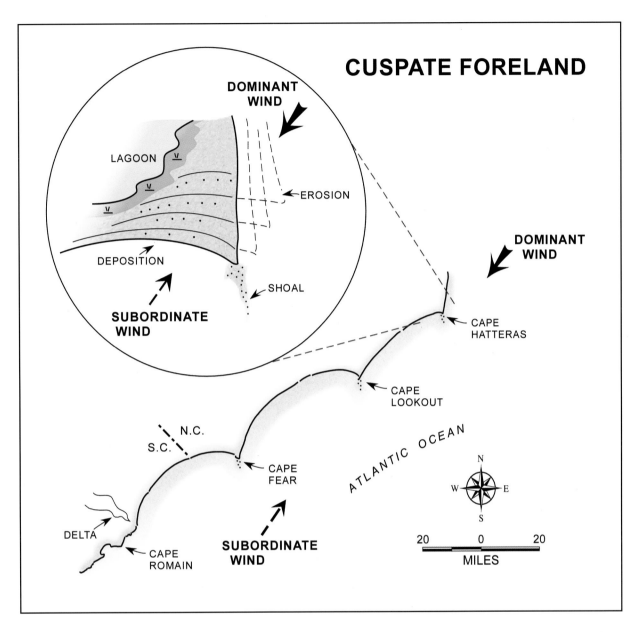

FIGURE 108. Typical configuration of the cuspate forelands of North Carolina (called capes), showing the importance of two opposing wind directions. In the case of the North Carolina capes, the northeast wind is the dominant one, creating waves that erode the northern flank of the headlands.

topography (discussed earlier; Figure 22C) make a bulge in the shoreline as the more exposed side of the sand bulge orients perpendicular to the wave approach direction (parallel with the approaching wave crests). Where that reorientation takes place, the rate of sediment transport slows down because of the near-parallel wave approach direction; however, the bulge continues to grow slowly further

out into deeper water. When the wind blows from the opposite direction, the other side of the sand bulge responds in a similar fashion, creating an arrow-shaped tip of the bulge that builds out into even deeper water. The net result is a series of **spits** (small features in a water body with limited fetch), or **forelands** (larger features in water bodies with essentially unlimited fetch), with relatively

171

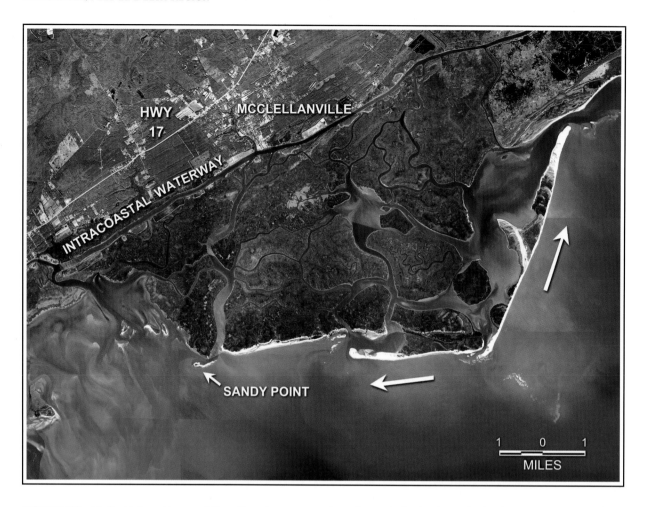

FIGURE 109. Vertical infrared image of Cape Romain, an arrow-shaped cuspate foreland, acquired in 2006, courtesy of SCDNR. Each flank of the cape is made up of a landward-migrating (transgressive) washover terrace system that terminates in a recurved spit complex. Arrows indicate that longshore sediment transport is to the north on the northeast flank of the foreland and to the west on the southwest flank. The landward-migrating islands of the foreland are presently retreating at the rate of several tens of yards per year.

equal spacing between them. With the strongest longshore sediment transport, and even erosion in some cases, occurring in the bays between the original spits, sediment is added to the spit form, which continues to project offshore. A splendid example of this continued growth of the spit ends was observed by Zenkovitch on the Chukchi coast of Russia, where the growing ends of such spits on opposite sides of a lagoon eventually met, causing the formation of a number of circular bays where the open lagoon formerly existed (see Figure 110). We have seen similar circular bays in former lagoons on the Alaskan coast of the Arctic Ocean,

as well as in Nantucket Harbor, Massachusetts. The larger the waves, the greater the distance required for the transport system to become "saturated," hence the large spacing of the cuspate forelands on the North Carolina coast (80/90 miles). When the winds from the opposite directions are equal, the spit form is very symmetrical, but if one of the winds dominate, the form is more asymmetrical and is prone to migrate in the downdrift direction of the larger waves, as illustrated in Figure 108.

Also similar to Carolina Bays, a number of exotic origins have been proposed for the cuspate forelands on the North Carolina/South Carolina

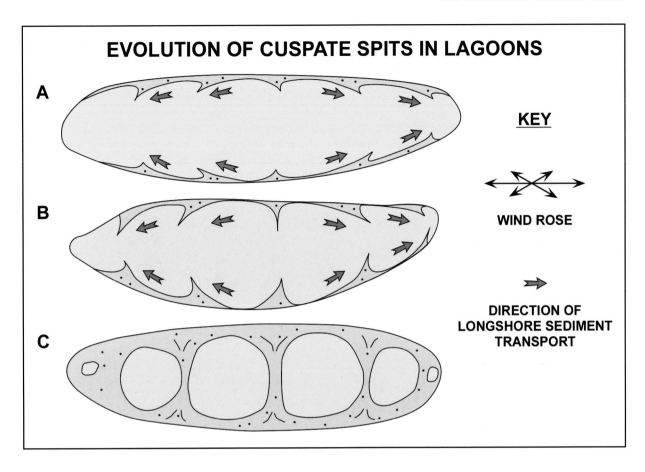

FIGURE 110. General model of how the growing ends of cuspate spits in a lagoon might eventually meet, causing the formation of circular bays. Based on observations made by Zenkovitch (1967) on the Chukchi Sea coast of Russia.

coasts, such as eddies off the Gulf Stream, location of river outlets, and positioning over ancient delta systems. Again, following the dictates of Accam's Razor, the wind-related cause-and-effect origin is the simplest and most logical (and most easily proved) of the different theories (Rosen, 1975; another Hayes student). In the case of the North Carolina capes, the two wind directions are:

1) The most **dominant wind** that blows from the northeast during extratropical cyclones (northeasters)
2) A **subordinate wind** from the southwest that blows around the Bermuda High mostly during the summer months.

As shown by the sketch in Figure 108, the waves generated by the northeasters (as well as by swell

from offshore storms and the occasional hurricane) tend to erode the north flank of the North Carolina capes, whereas the weaker southwest waves mold the eroded sand to the southern flank of the cape. As a result, the cape "migrates" ever so slowly to the south. This general model fits the three North Carolina capes perfectly.

The outer shore of Cape Romain is one of the most erosional coastlines in the state. A study by our group published in 1975 showed that, except for the outbuilding recurved spits at each end of the cape, the shoreline retreated about 20 feet per year on the average between 1941-1973. Based on later observations, this rate of retreat has increased. The photograph in Figure 111, taken shortly after hurricane *Hugo*, showed that the beach retreated significantly during the storm, creating a new tidal inlet. This "new" inlet is now used by shrimp boats

FIGURE 111. View of the southwest end of Cape Romain (Sandy Point) a few days after hurricane *Hugo* (11 October 1989). The white band of sediment, a mixture of sand and shell, is a washover terrace that advanced landward several 10s of feet during the storm. The dark layer seaward of the washover terrace is exposed backbarrier muddy sediments. As a result of the landward migration of this barrier island, a new tidal inlet was created where it intersected the large tidal channel in the foreground.

operating out of the town of McClellanville. The accelerated rate of erosion of the Cape in recent years is probably due, at least in part, to the diminished sand supply to the Santee River delta as a result of the building of the dam on the Santee River in 1942. The Santee Delta front is located only about 15 miles to the north of the arrowhead point of the cape. Except for the prograding northern flank of the cape, the barrier islands in this area have some of the highest rates of landward migration in the state (Ruby, 1981). The typical morphology of the landward-migrating barrier islands of the Cape is shown in Figure 112.

Cape Romain does not conform to the general morphological model given in Figure 108. As

shown by the recent image in Figure 109, the two flanks of the cape are relatively symmetrical. Also, the northern flank has shown significant longshore sediment transport to the north (a major exception to the dominant southerly transport on the state's coastline) for at least several hundred years (based on study of historical maps and charts).

We speculate that the reason this cape is different from the other three is related to two factors:

- Sheltering of the cape from the dominant northeasterly waves at the present time (to some extent) by the mass of the Santee Delta, as well as by the shoals offshore of the delta.
- The rapid retreat of the cape over the past few

174

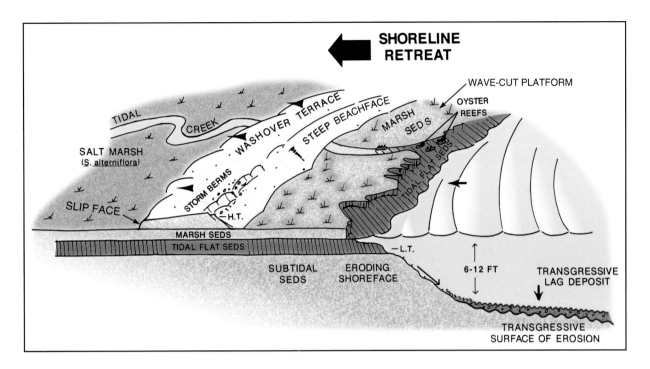

FIGURE 112. Low-tide morphology of the landward-migrating (transgressive) barrier island on the southwest end of Cape Romain (at Sandy Point; see photograph in Figure 111), based on observations made during the late 1990s. At that time, a wide wave-cut platform composed of salt-marsh sediments was exposed at low tide (also visible in the photograph in Figure 111). A scarp displaying a remarkable array of erosional features was eroded into the outer edge of the exposed salt-marsh sediments. As the shoreline retreated, tidal creeks and oyster reefs were exposed along the eroding platform. The beach face was steep, because of an abundance of coarse-grained shell material in the beach sediment, and the washover terrace was terminated in a 3-foot high slip face on top of the living marsh surface (this abrupt margin is also visible in Figure 111). Steep storm berms composed mostly of oyster shells were present in places along the top of the beach face. According to Walter J. Sexton, this landward-migrating barrier island at Sandy Point was nearly gone in 2007, the sediments being deposited into the deep tidal channel on its landward side that is shown in Figure 111.

hundred years conceivably moved the present form further into the shadow of the delta than it was when it first started to develop. In the early years, it may have had the same morphology as the other capes, because it was more exposed. The bathymetry of the shelf off the present cape shows that complicated shoals now occupy the space where the cape was formerly located (Figure 90B).

Places to Visit

About the only place to get to this shoreline is from the charming, historic town of McClellanville, a town mostly occupied by locals, notably shrimp fishermen. The movie *Paradise* (1991), starring Don Johnson and Melanie Griffith, was filmed in this town. Would you believe Don played a shrimp-boat captain?

Though few, if any, people died in this town during the storm, there were some real horror stories related to hurricane *Hugo*, which had its maximum storm surge of over 20 feet in this area. For example, a number of people had retreated to the local school building to escape the storm, some of them standing on tables to avoid the rising water. When the governor's helicopter approached the school the next morning (the storm passed over around midnight), the peoples' cars were all in a big pile on one side of the school. Some of the more formidable shrimp fishermen rode out the storm in their boats, but many of them said they "wouldn't do it again," a refrain heard from most people who chose to remain put during the storm, no matter where they were located along the coast. Looking at the town now, you would never guess that it had been flooded by a hurricane.

We highly recommend you take a boat tour from McClellanville out to the southernmost limb of Cape Romain, should one be available. We have made the trip to the Sandy Point spit, which lately has been eroding at a rate of around 100 feet per year, probably around a hundred times during our continuing education seminars. We have been told that the barrier island south of the new inlet (Figure 111) has now been nearly washed away. However, those beaches that are left on this part of the Cape are probably the best place in the state for shell collection and the shorebirds are unbelievably abundant during migration. Also, it is a great boat ride through the very extensive marsh. A word of advice: Don't make this trip if a thunderstorm is predicted, speaking from experience!!! We have also been told, however, that it is very difficult to get permission to walk on the beaches of this part of the Cape these days because of the high sensitivity of the area for bird nesting, etc. Good luck!

Going south from McClellanville on US 17 for about 2.5 miles you encounter on the left the **Tibwin Nature Trail**, which is located in a part of the **Francis Marion National Forest**. At first, this trail is rather mundane until you get down to the freshwater ponds. On the ocean side of the ponds, you can walk along beside a wide salt marsh with excellent examples of the marsh vegetation. Numerous waterfowl and the occasional river otter inhabit the ponds. You are not supposed to walk around the main pond between 10 November and 10 March. From the end of the trail, you can see across to the Intracoastal Waterway, an appealing view.

BULLS BAY

It is impossible to explore Bulls Bay (Figure 113) without a watercraft and near impossible with one. Why this open bay exists along this segment of the coast that is typically made up of barrier islands is still somewhat of a mystery to us. We speculate, based on limited evidence, that a lowstand valley, formed during the Kansan or Illinois glaciation, may underly the bay (D.J. Colquhoun, pers. comm.), but we have not done the geophysical surveys necessary to prove or disprove this theory. If true, the valley may not have filled in with sufficient sediment to elevate the bay bottom to a level adequate for barrier islands to form. The large lobe of sand on the shelf just off the bay, shown in Figure 90A, is probably the remnant of an ancient delta of the Santee River. This is another line of evidence supporting the idea that a lowstand valley underlies the Bay, the delta having been formed by the river that carved the old valley. As shown by the low-tide image in Figure 75B, the bay is presently filling in with sediments as evidenced by the extensive tidal flats and tidal channels. A small island on the northwest side of the bay, which is partially submerged during high spring tides, appeared to be near evolving into a barrier island in the 1980s, but it was planed off during hurricane *Hugo*. This embryonic island is making a comeback at the present time (Figures 75B and 113).

There are two sites worth a visit within this subcompartment that you can drive to:

1) Buck Hall Recreational Area
2) The Sewee Visitor and Educational Center

The **Buck Hall** landing is located about 6 miles south of McClellanville on US 17. This Recreational Area, which contains a boat ramp, picnic area, and restrooms, is positioned on a Pleistocene highland on the bank of the Intracoastal Waterway. This is a good spot to have a picnic while overlooking the wide salt marsh that borders Bulls Bay.

Continuing south on US 17 for about another 8.5 miles further, you come to the **Sewee Visitor and Environmental Education Center,** which has some instructive exhibits and useful reference material. This center is definitely worth a stop.

FIGURE 113. Vertical infrared image of Bulls Bay acquired on 11 May 2000, courtesy of Earth Science Data Interface at the Global Land Cover Facility. Arrow A points to what appears to be an embryonic barrier island (Bird Island) and arrow B is in approximately the same location from which the oblique aerial photograph in Figure 111 was taken. In the eleven years between the date of hurricane *Hugo* (September 1989) and the time this image was taken (May 2000), the shoreline of the southwest end of Cape Romain (vicinity of arrow B) had continued to retreat in a dramatic fashion.

13 THE BARRIER ISLANDS

INTRODUCTION

A 45-mile stretch of barrier islands make up the shoreline between Bulls Bay and St. Helena Sound, the boundaries of this major compartment (Figure 114). Three of these, Bull, Isle of Palms, and Kiawah Islands, are classic prograding, drumstick-shaped barrier islands. In fact, these islands are the ones about which many of the original concepts on the morphology of mesotidal (mixed energy) barrier islands, as well as their stratigraphic history, were developed (Hayes, 1977; 1979; Fitzgerald, 1977; Hubbard, 1977; Moslow, 1980). A recent paper by Harris, et al. (2005), which also discusses the classic nature of these barrier islands, emphasizes the importance of the underlying geological framework. Except for Bull, Capers, Morris, and the landward-migrating barrier islands north of Edisto Beach State Park, they are all developed for human habitation, at least to some extent, with four being notable tourist attractions. There are also several river systems in the compartment, many with historic plantations along their banks. For purposes of discussion, we have subdivided this compartment into five subcompartments:

- **Bull Island area** – These three barrier islands, Bull, Capers, and Dewees, which can be reached only by boat, are all prograding in a geologic sense, with Bull exhibiting the classic drumstick shape. Bull Island is included in the Cape Romain National Wildlife Refuge and Capers Island is a state park. Dewees, which has been developed rather rapidly over the past few years, now contains 40+ dwellings.

- **Isle of Palms/Sullivans Island** – Two very distinctly prograding barrier islands are located just to the north of the entrance to Charleston Harbor. These islands, which have been settled for many decades, have ready access across two causeways from the mainland. An upscale, gated community, Wild Dunes, is located on the north end of the Isle of Palms.

- **Folly Beach/Morris Island area** – Folly Beach, a beach-front town with local flavor, is on Folly Island, formerly known as Coffin Island. Folly Island has a complex erosional/depositional history, with erosion dominating in the last several decades. This subcompartment also includes Morris Island, an extremely erosional, unpopulated island located between Folly Island and Charleston Harbor.

- **Kiawah and Seabrook Islands** – Both of these islands are predominantly gated, upscale developments, with general public access only being available at Beachwalker Park, which has one of the most beautiful natural beaches in the state.

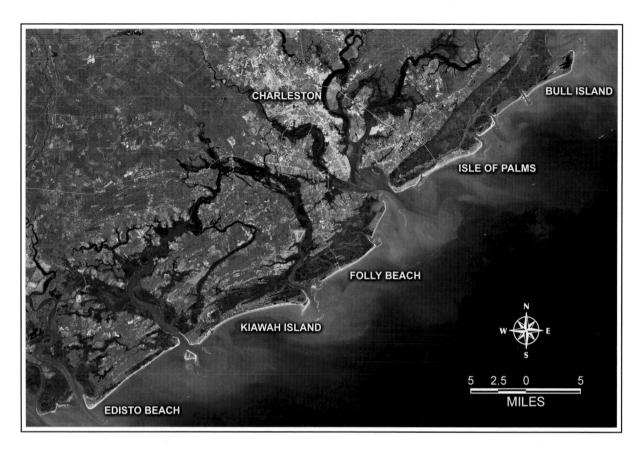

FIGURE 114. Vertical infrared image of the Barrier Islands compartment acquired on 23 October 1999, courtesy of Earth Science Data Interface at the Global Land Cover Facility.

- **Edisto Beach area** – The Edisto Beach State Park, with its relatively large waves and ample camping facilities, is one of the most popular parks in the state. The park is on the updrift limb of a large recurved spit that historically has built to the south into St. Helena Sound. Presently, the beaches in the park and those to the north of it are eroding rather dramatically, because of a major shift in the main ebb channel of the huge ebb-tidal delta of the North Edisto Inlet, located approximately 7 miles to the northeast of the park.

BULL ISLAND AREA

Morphology and Origin

Examination of the image of the area shown in Figure 115 reveals several significant contrasting features with regard to the morphology of this subcompartment:

- The linear contact between the higher Pleistocene mainland, with an age measured in 10s of thousands of years (shown in red), and the lower marsh complex (in gray), which has formed over the past 4,500 years or so.
- The Intracoastal Waterway, which cuts across the marsh system.
- The parallel, prograding beach ridges on each of the islands.
- The three natural, unmodified (by man) tidal inlets, with their conspicuous ebb-tidal deltas, Price, Capers, and Dewees, the last two of which have striking downdrift offsets (see image in Figure 30 and the oblique aerial photographs in Figures 31 and 116, which illustrate the offsets).

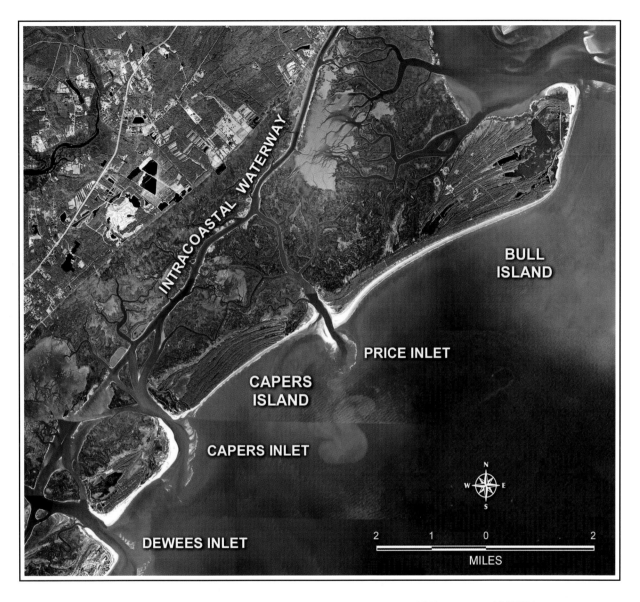

FIGURE 115. Vertical infrared image of the Bull Island subcompartment acquired in 2006, courtesy of SCDNR.

• The drumstick configuration of Bull Island.

This is an area where Hayes and his graduate students at the University of South Carolina conducted a considerable amount of research in the 1970s and 1980s. For example, a study by Tim Kana (Kana, 1979) established the longshore sediment transport rate of approximately 150,000 cubic yards per year (to the south) on Capers Island, and Duncan Fitzgerald's dissertation on the evolution of Price Inlet (Fitzgerald, 1977) remains today as one of the classic studies of tidal inlets. There were

other studies of note as well (e.g., Tye, 1981; 1984; Knoth and Nummedal, 1977).

One of the questions that puzzled us at first was where did the sand come from to build these islands? This was particularly odd because of the fact that the open bay to the north (Bulls Bay; Figure 113) would not allow longshore sediment transport from the north along a beach zone. Furthermore, there is no evidence for longshore sediment transport from south to north in this area. Studies by Griffin (1981) and Sexton (1987) provided the most likely answer to this question.

FIGURE 116. The Bull Island subcompartment. (A) Oblique infrared photograph, the bottom of which shows the northeast end of Bull Island. The top of the photo marks the southwest end of the Isle of Palms. Compare this view with the images in Figures 30 and 115. Photograph by Walter J. Sexton taken at low tide on 23 November 1981. (B) True color view of the same area as is shown in A, but at a lower altitude. Note the well developed intertidal bar system exposed at low tide. The white sand in the vegetation behind the beach (arrows) is dune sand deposited by hurricane *Hugo* (1989). Photograph taken at low tide on 5 May 1997.

As noted in the earlier discussion of Figure 90A, there is a huge lobe of sediment on the continental shelf just off the entrance to Bulls Bay. Sexton (pers. comm.) confirms that the sediment in this lobe has a riverine character, that is, its mineralogy and texture indicates that it was brought there directly by an ancestral Santee River. Martha Griffin's study of the mineralogy of the sand on the beaches of the islands themselves (Bull, Capers, Dewees, and Isle of Palms) confirms their origin from that same riverine source, because of the elevated feldspar content of the sand. Orthoclase ($KAlSi_3O_8$), with its less chemically stable elements K and Al, is one of the more common feldspars present. Most of the other beaches in the state are dominated by quartz (SiO_2), a mineral more resistant than feldspar to chemical weathering and abrasion. Based on this confirming evidence, we conclude that the sediments in this whole barrier island chain, from Bull Island to the entrance to Charleston Harbor, were derived mainly from this offshore sediment source (abandoned delta lobe off Bulls Bay).

Places to Visit

Inasmuch as the barrier islands in this area

can be reached only by boat, there are just two possible ways to visit them, unless you go by means of your own watercraft. **Price Inlet** is a very popular boating destination, with the boats usually departing from the Isle of Palms Marina, located on the Intracoastal Waterway near the north end of the Isle of Palms. From there, you could take either your own watercraft or join a tour (check the internet for possibilities). You can also get to the inlet by launching your boat at Garris Landing, which is discussed below. If you do find a way to get to Price Inlet, there are many great activities to pursue (we recommend that you make such visits at low tide, with low spring tides being the best):

- The fine-grained sand beaches both north and south of the inlet make for great walks in a totally natural setting (see oblique aerial photograph of Price Inlet in Figure 38B).
- Walk south along Capers Island for about 3/4 mile and you come to an eroding maritime forest in the surf. Great photography!!! Stratigraphic studies by Tye (1981; 1984) show that Capers Island has historically been a prograding barrier island, starting with a preserved, previously landward-migrating element on the landward side of the island around 4,500 years ago, followed by seaward progradation of a series of beach ridges, as is shown by the model given in Figure 33. However, the most seaward of the prograding beach ridges have been eroding in a rather spectacular fashion for at least the past 50 years or so. Two possible reasons account for this, either: 1) an unusual amount of trapping of sand in the longshore transport system by the ebb-tidal delta of Price Inlet; or 2) a general decrease in the supply of sediment from the offshore delta lobe discussed earlier.
- At about 30 yards north of the northern edge of the trees, you will be standing exactly on the spot where the main channel of the inlet was located in 1822 (Figure 117). Studies by Bo

Tye (1981; 1984) showed that the inlet tends to migrate very slowly to the southwest until such time that a hurricane cuts through the new land developed in the lee of the migrating inlet. The new cut usually occurs at a spot near where the inlet is now located (2007). We are not sure exactly when this natural relocation of the inlet last occurred, but stratigraphic evidence clearly shows that it has happened more than once in the past 4,500 years. It is important to realize that the inlet does not migrate back and forth, up and down, the beach. Rather, after slowly migrating to the southwest for at least several decades, it shifts abruptly back to the northeast as the ebbing flow of the storm surge generated by a hurricane cuts a new channel directly across the beach.

- This inlet expanded during hurricane *Hugo* (1989) to the extent that most of the foredunes between the inlet throat and the trees to the south were eroded away. The line of foredunes that you will probably see on your walk down the island was rebuilt by normal processes within a few years after the hurricane. This beach does experience striking changes in erosion and deposition under natural conditions, so we cannot predict exactly how it will look at any given time. We have continued to visit this inlet over the years since 1973, and it continues to change in a remarkable fashion.
- At low tide, you can walk on the many features of the ebb-tidal delta model presented in Figure 37 and shown in the oblique aerial photographs of Price Inlet in Figure 38B and 117, including swash bars, marginal flood channels, and channel-margin linear bars. This is a spectacular spot to observe these features, and we have taken well over 100 training seminars to this spot over the years because it is so classic in its many aspects.
- Pretty good shell collecting, especially on the Capers side.
- Birding is also good. Spectacular flights of

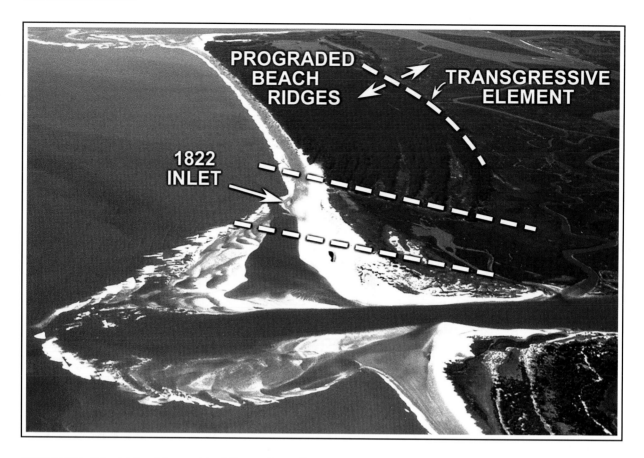

FIGURE 117. Price Inlet and Capers Island. The position of Price inlet in 1822 is shown. Note the truncated beach ridges at north end of Capers Island (middle distance; see also Figures 115 and 116). These ridges were truncated by the tidal inlet when it was at that position in 1822. Since then, the inlet has reoriented to the north and the old channel has become filled in with fine-grained sediments, creating a clay plug in the old channel. The stratigraphy of this section was studied by R.S. Tye (1981; 1984). Photograph taken at low tide in the winter of 1975.

hundreds of black skimmers and shorebirds come to mind.

There is a ferry to **Bull Island**. We took the following ferry schedule off the internet. This might change. You should check first. Fee is $30 for adults. A bargain!!

March 1 - November 30
Tues., Thurs., Fri., and Sat.
Depart at 9 am & 12:30 pm
Leave Bull Island at 12 & 4 pm
December 1 - February 29
Sat. only - Depart at 10 am; Leave Bull Island at 3 pm

How to get to the ferry? From Charleston, go north on US 17 for about 14 miles and on the left you will pass a wide expanse of black needlerush (*Juncus roemerianus*) marsh in the headwaters of the Wando River estuary. A little further along, you reach the Sewee Restaurant, where you take a right turn on the road to the Bull Island Ferry. Once you make the turn, after a drive of about 5 miles and by following the signs, you reach what is now called **Garris Landing**. This is where you catch the ferry to the island. Want to camp over there on the island? Go right ahead, we've tried that. Mosquitos!!! Once on the island, there are several nice trails across this completely undeveloped prograding barrier island. But, don't walk in any tall grass. The worst case of chiggers we have ever had were the result of doing the same. We really don't want to discourage you from making this trip or anything, but just be careful!!

While out there, you will be walking around on a part of the **Cape Romain National Wildlife Refuge**, one of the largest on the east coast. Some facts about this refuge (from the refuge web page):

- It consists of 66,267 acres, 28,000 of which are preserved with the National Wilderness Preservation System.
- It ranges for 20 miles along the coast and contains 36,267 acres of a variety of coastal environments and 30,000 acres of open water.
- Over 227 species of birds can be found on the refuge.
- It has the largest nesting rookery for brown pelicans, terns, and gulls on the South Carolina coast, as well as the largest nesting population of loggerhead sea turtles outside the State of Florida.
- At one time, five red wolves were placed on the island, but even though they did survive hurricane *Hugo* (1989), we don't know what their status is at this time (2007).

Most of the trees on Bull Island were blown down during hurricane *Hugo*. There was some beach erosion in places, but the island appears to be completely recovered from the storm by now (2007), at least in the southern portion. Historically, the island has been erosional on the northern end and accretionary on the southern end. Between 1934 and 1994, the northernmost part of the island eroded (lost sand) at the rate of about 40 cubic yards per linear yard of beach per year, and the southernmost segment gained about 12 cubic yards per linear yard of beach per year (Kana and Gaudiano, 2001).

Even if you don't go out to the island, a visit to Garris Landing would be interesting. This very popular boat landing is a great launch site for canoe and kayak trips as well. From there, you can paddle along extensive tidal channels through the marsh, see a multitude of birds, and so on. Also, you could come out at low tide and walk to the end of the pier to view shorebirds, waders, and other avifauna.

If you are interested in sampling the locally harvested coastal Carolina sea food, we recommend the Sewee Restaurant, which is located on US 17 at the turn to the Bull Island Ferry. Also, as you drive south along US 17 after your feast at the Sewee Restaurant (Note: We do not own any part of this business, unfortunately), you will see a number of ladies marketing hand-made sweetgrass baskets, using a handicraft handed down from West Africa.

The following information on this form of artistry is taken from www.hort.purdue.edu.

"Sweetgrass (Muhlenbergia sp.) is a native, perennial, warm-season grass found growing sparsely in the coastal dunes extending from North Carolina to Texas. The "threads" or long, narrow leaf blades of this grass have been harvested by direct descendents of enslaved Africans of antebellum South Carolina and used as the principle foundation material for constructing African coiled basketry in the Southeast, especially near Charleston.
"The technique of basketmaking crossed the Atlantic in slave ships and took root in the new land. The art of making sweetgrass baskets is a three-century old African-American tradition that has been "passed-down" over the centuries from parent to offspring. Sweetgrass basketmaking is one of the earliest traditional crafts with a rich documented history of "carryover" from African enslavement and plantation days to the present (McKissick Museum 1988). Although the materials used are different in the United States, the form and function of the African counterparts of sweetgrass baskets are unchanged to this day (Mary Jackson pers. comm.) .... Sweetgrass grows on many of the barrier islands along South Carolina's coast, such as Kiawah, Seabrook, Dewees, Bull, Fripp, and Hilton Head."

ISLE OF PALMS/SULLIVANS ISLAND

Morphology

Being prograding barrier islands, these two islands have not undergone serious beach erosion like that experienced by its neighbor to the south, Folly Island. The Isle of Palms has the characteristic drumstick shape typical of most of the larger prograding barrier islands in the state (see vertical image in Figure 118 and oblique aerial photographs in Figure 119). As noted earlier, there were some spectacular impacts of hurricane *Hugo* on both of the islands, particularly where the two associated tidal inlets, Dewees to the north of the Isle of Palms, and Breach Inlet, expanded during the storm. The central part of the Isle of Palms was also eroded in a spectacular fashion, with several houses being

completely removed from their foundations and moved into either the neighbor's front door or down some of the side streets for more than a block (see Figures 61 and 62). All of the damaged houses along the front row of both islands have long been replaced or repaired, with some being considerably larger than the original. As noted earlier, however, the actual loss of land upon which the houses were originally built was minimal, because of the accretional nature of these islands.

From time to time, there are some relatively minor erosional episodes on the island, especially in areas located close to the large Dewees Inlet on the north side of the Isle of Palms. Detachments of major swash bars from the large ebb-tidal delta at Dewees Inlet have resulted in episodes of attachment of these bars to the beach of the Wild Dunes development (see Figure 119B) similar to

FIGURE 118. Vertical infrared image of the Isle of Palms/Sullivans Island subcompartment. Acquired in 2006. Courtesy of SCDNR.

FIGURE 119. The mesotidal barrier island system to the immediate north of Charleston. (A) Oblique view looking northeast in April 1975. Sullivans Island is in the foreground. Note drumstick shape of the Isle of Palms, and the downdrift offset at Breach Inlet (compare with Figure 118). (B) Huge intertidal bar welding onto the beach on the northeast end of the Isle of Palms (arrow). Compare with Stage 2 in Kana's model given in Figure 56B. Photograph taken at low tide on 5 May 1997.

the ones illustrated at Stono Inlet in Figure 56A. Over the years, Tim Kana has conducted research on these events, which he terms *"the process of shoal by-passing – the episodic transfer of sediment from inlet or delta shoals to a beach."* These "episodic transfers" create a cycle of erosion and accretion at the north end of some barrier islands in the state, most notably Kiawah and the Isle of Palms. The model for this process, which involves three stages, is given in Figure 56B (from Kana et al., 1999). The three stages are:

Stage 1. An offshore shoal "detaches" from the swash platform of the ebb-tidal delta (see ebb-tidal delta model in Figure 37), becomes isolated, and is eventually driven landward by the waves.

Stage 2. The shoal, which typically has a horseshoe shape, attaches to the beach at one or both limbs of the "horseshoe."

Stage 3. As the main body of the shoal welds to the beach, the waves disperse the sand in either direction away from the point of detachment.

Fairly early in Stage 1, parts of the beach not exactly in the lee of the shoal, where sand tends to accumulate in a triangle-shaped projection called a tombolo (Figure 21A), may become erosional to the point of causing concern for the land owners. Waves refracting around the bar set up longshore transport of sediment toward the tombolo (Stage 1 in Figure 56B). An example of one of these shoals in Stage 2 on the northern shore of the Isle of Palms in 1997 is shown in Figure 119B. Notice how the two limbs of the "horseshoe" have wrapped around the tombolo that originally formed in the lee of the bar. Obviously, after the bar has welded to the beach and the sand is dispersed along shore, the beach in that area will be built out, at least in the short term.

History

Presumably because of their proximity to historic Charleston, these two islands have a long history of human activities:

- Fort Moultrie was established near the south end of Sullivans Island, and it was there that the first rebel victory during the First Civil War took place on 28 June 1776. The famous "Swamp Fox" and General Moultrie, both experienced Indian fighters, were part of the crew stationed at the Fort.
- In that same war, a British Army contingent of 2,500 men who had landed on the Isle of Palms attempted to raid a colonial encampment on adjacent Sullivans Island, but failed because so many men drowned attempting to cross the treacherous waters of Breach Inlet (Figure 118). Heck, they don't have barrier islands and tidal inlets in Great Britain.
- During the Second Civil War, this area was a point of departure for the *CSS Hunley*, the first submarine to sink an enemy vessel. Unfortunately, or fortunately, depending on your point of view, it was soon lost at sea as well. The wreck of the *Hunley* was recently discovered offshore of Sullivans Island and is being restored.
- As early as 1906, a 50-room resort hotel was built on the Isle of Palms.
- Development of permanent residences on the two islands began in earnest at the end of World War II. Speaking of which, if you visit Patriot's Point, don't miss a tour of the aircraft carrier *USS YORKTOWN*, an engrossing historic experience.

- In the 1970s, the rest of the world discovered Sullivans Island and the Isle of Palms, and real estate development blossomed.
- Since about the early 1980s, the gated community at the north end of the island, Wild Dunes, with its world-class golf course and excellent restaurants, has become one of the finest tourist destinations of the mid-South Carolina coast (no, we don't own anything there either).

Places to Visit

The first place we will discuss is **Pitt Street bridge**, one of our favorite birding spots in the whole state. Make sure you go at low tide! To go there from downtown Charleston, drive north on US 17 and after you cross the fabulous new Ravenel bridge over the harbor, take state road 703N in the direction of Sullivans Island. After a few minutes, you will cross Shem Creek, with its shrimp boat docks and a number of associated seafood restaurants. During hurricane *Hugo* (1989), a couple of our favorite restaurants that used to be on the bank of the creek, as well as the bridge across the Intracoastal Waterway, were washed away. After you cross the bridge, keep going straight on Whildon Street, driving into historic Old Mount Pleasant. Keep going until the road's name changes to Royal Street. A couple of hundred yards further along from there, take a right and drive one block to Pitt Street, where you turn left and drive to the end of the road. Even if you weren't heading to one of the best birding spots in the state, this historic district, which survived *Hugo* obviously, would be worth the visit. At the end of the road, the southwest end of the old bridge that used to cross to Sullivans Island and its surrounding area has been converted into a park, called the **Pickett Recreation Area**.

Walking at a higher elevation along the roadway leading to the old bridge, which is no longer in place, you have a great view looking down on several tidal channels, extensive tidal flats, and salt marshes (if you come at low tide, of course). We have birded here many times during our visits with Michel's family in the Charleston area, and we were never disappointed.

On 13 January 2007, we were there for about an hour or so and we saw the following birds: snowy egret, a group of 10 hooded mergansers feeding (see photograph in Figure 120), ring-billed gull, boat-tailed grackle, several small shorebirds (sandpipers, dunlins, etc.), several marbled godwits, double-crested cormorants, brown pelicans, willet, tricolored heron, and great blue heron. There are some good views of historic Charleston from this area as well. If you are into birding, we would definitely recommend a couple of hours during low tide at this site. Don't be surprised if you meet other birders there, especially on the weekend.

To get to the next recommended site, **Sullivans Island,** go back out Pitt Street and turn right on Center Street, which puts you back at state road 703. From there, turn right and drive across the salt marsh (*Spartina alterniflora),* as expected in an area thus far removed from a fresh water source, to Sullivans Island. Continue until you cross a drawbridge across the Intracoastal Waterway. For the record, all of the bridges to Sullivans Island and the Isle of Palms were severely damaged during *Hugo.* After crossing another stretch of marsh, you arrive on Sullivans Island. At the stop sign you could take a right, if you are inclined to see an historic fort. **Fort Moultrie** was named for a prominent general and notable Indian fighter, who fought for the rebels in the First Civil War. It is 1.2 miles from the turn to the Fort. As noted earlier, this island and

FIGURE 120. Hooded mergansers at the Pickett Recreation Area in Mt. Pleasant. Photograph taken on 13 January 2007.

the Isle of Palms were settled decades ago. In the early post World War II days, the residences were owned mostly by local Charlestonians, who had summer houses there. In fact, Michel, who grew up in the Mount Pleasant area, spent some of her childhood in a family house located on Sullivans Island. As we were doing a tour of this area while preparing this book, Hayes couldn't concentrate on what he was doing for her constantly pointing out places where she played as a kid, crawling through dungeons at the Fort and so on.

Most of these houses on the road down to the Fort survived *Hugo*, but there was some pretty serious wind damage in places. You can get to the beach in this area by driving to a parking area at the end of the road. However, there are some shore protection structures, seawalls, riprap, etc., on the beach there. This is probably not a very good place to swim, anyway, because of the potential for strong tidal currents running into Charleston Harbor.

There is a parking space and beach access to the open ocean at the **Sullivans Island Light**. Also, it is common for these older developments, like here and Folly Beach, to have designated places where you can walk by the cottages and get to the beach. Parking is pretty sparse, however. Nice beach though, wide and flat, on this prograding barrier island.

Driving north, you will eventually reach **Breach Inlet**, which separates Sullivans Island from the **Isle of Palms**. The bridge over the inlet was partially washed out during *Hugo*, and some of the houses south of the inlet were severely impacted as the inlet expanded in response to a storm surge of around 15 feet at this location (see Figure 67 for an illustration of this type of expansion).

There is a parking area at the north side of the bridge, from which you can walk out along the beach by the main channel of the inlet. There is usually a major intertidal bar on the beach at this location that you can walk to at low tide. Tidal currents running in and out of the inlet can be very strong at times, so be cautious. Remember the fate of the poor British soldiers during the First Civil War.

Driving north from the inlet, at about 1.2 miles, the road intersects with a causeway (route 517) that was built across the marsh after *Hugo*. There are some places where you can park and get beach access in this area as well. A few hundred yards ahead, the main road, which is blocked by a huge church, continues to the right at first and eventually swings back to the north (left) on the landward side of the front-row beach houses. At the present time (January 2007), the beach appears to have built out several 10s of yards since *Hugo* at this location. The front-line cottages along this stretch were hit pretty hard during *Hugo*, some of them being washed landward down the streets for a block or two. During the hurricane, the water came through and under the beachfront houses and filled up this street. Needless to say, most of the new McMansions that you see along here were built after 1989. In this upscale area, there are fewer beach access points, not nearly as visitor friendly as Sullivans Island or Folly Beach. The square footages of the houses along this stretch have exploded in the last 10 or 15 years, this beachfront formerly being occupied by much smaller cottages built decades ago.

A bit further along, you reach the entrance to Wild Dunes, a gated community, which has limited access. If so inclined, you may be able to gain entrance to play golf on its famous golf course, which winds through some high dunes.

Further down this same road by the Intracoastal Waterway is a first-class marina, where watercraft can be launched for a fee. That marina was devastated during *Hugo*, with many of the boats being stacked up in a big pile on the landward side of the Intracoastal Waterway.

Driving back along the road toward the causeway, you will probably be impressed with the size of some of these houses, but "you ain't seen nothing yet" until you have driven to the north end of Kiawah Island, where the McMansions are global in scale. We have been coming to the Isle of Palms routinely for the past 30 years or so, and the changes have been phenomenal. Of course, *Hugo* caused the biggest changes, and after that, the changes in the scale of the development and the increased real estate prices in the past 10 years or so have gone off the charts. The eye of *Hugo* passed right over this area a little after midnight on 21 September 1989, over 17 years ago, and us locals still talk about it like it was last week.

Driving back across the northern causeway as you are leaving the Isle of Palms allows a great view of the wide expanse of salt marsh in this backbarrier area. As you leave the marsh and drive up onto the mainland, you will be driving on Pleistocene sediments deposited about 100,000 years ago, which is over an order of magnitude more than 6,000 years ago, just in case you were wondering.

Continuing on state road 517 you will eventually reach US 17. If you drive about 2 miles north on US 17, you will come to the entrance to the **Boone Hall Plantation**, which has been the site for several movies on the antebellum south. It has a noteworthy avenue of oaks and, if you are into gardening, you will probably find the formal garden to be an interesting and pleasant one to visit. Spring time is best, of course.

FOLLY BEACH/MORRIS ISLAND AREA

Morphology, Origin, and History

The morphology of Folly Island and Morris Island (Figure 121) is closely tied to the construction of the Charleston Harbor jetties in the late 1800s. Before the jetties were constructed, historic charts indicate that both islands were prograding, with an origin similar to the other islands of this type on the coast. That is, they were initiated as landward-migrating barrier islands around 4,500 years ago that eventually began to prograde as sand in the longshore transport system moved around the huge ebb-tidal delta off the entrance to the harbor. As the chart prepared by the British military in 1779

shows (see Figure 122), the main ebb channel of the ebb-tidal delta was deflected far to the south as a result of the strong northeast-to-southwest sediment transport system. Folly Island, which was called Coffin Island at that time, appears on the chart to have a drumstick configuration, a clue to its prograding nature.

Construction on the Charleston Harbor jetties was begun in the late 1870s and completed in 1898. These jetties were designed to reroute the main harbor channel to the southeast directly through the channel-margin shoals depicted on the 1779 chart (Figure 122). The goal was to prevent the shifting sand shoals associated with the ebb-tidal delta from obstructing navigation into the harbor. Some of the immediate and long-term effects of constructing the new jetties through the heart of the ebb-tidal delta, as well as the dredging of the new channel, follow:

- Obviously, the jetties met the original goal of improving navigation in and out of the harbor, a pretty amazing engineering feat considering how long they have been there.

- Sediment that normally moved to the south along the shore at a rate of approximately 150,000 cubic yards per year was either: 1) blocked and accreted on the northern islands; or 2) accumulated into the weir excavation by the north jetty to be later dredged and dumped far offshore at depths great enough to take that sand out of the normal longshore transport system. The jetties are almost three miles long.

- The extensive shoal off the main ebb channel of the ebb-tidal delta was no longer maintained in its position offshore by strong ebb currents that formerly flowed in that north-south oriented channel. Consequently, the exceptionally large volume of sand formerly part of the end of the massive ebb-tidal delta was moved landward by wave action where it accumulated on the beaches of Folly and Morris Islands. As a result, the shorelines of the islands accreted (grew

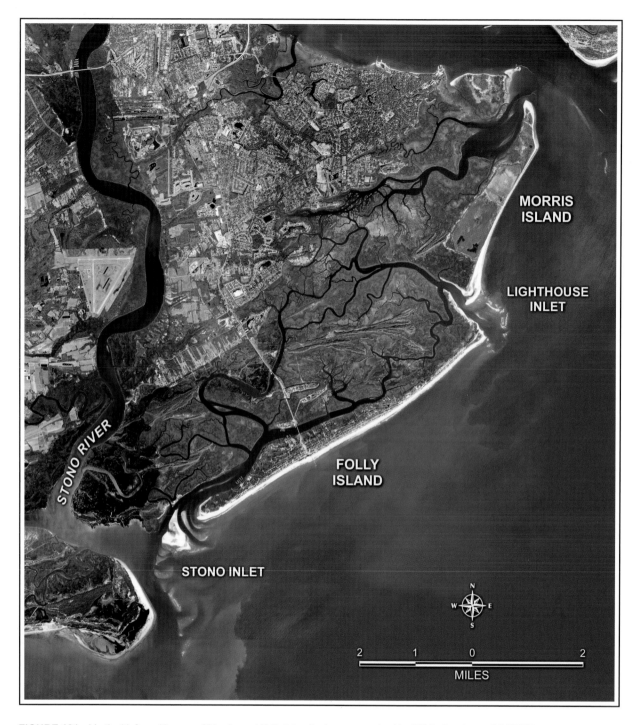

FIGURE 121. Vertical infrared image of Morris and Folly Islands. Image acquired in 2006. Courtesy of SCDNR.

seaward) more than 2,000 feet over the first few years after the jetties were built (Lennon, 2000). Such a process can take place naturally in some situations, but, in this case, that massive accretion was strictly man-influenced.

- By the 1930s, the sand initially deposited on the islands began to erode away, because no more sand was transported around the harbor entrance to the beaches to the south. Since that time, according to Lennon (2000), the

192

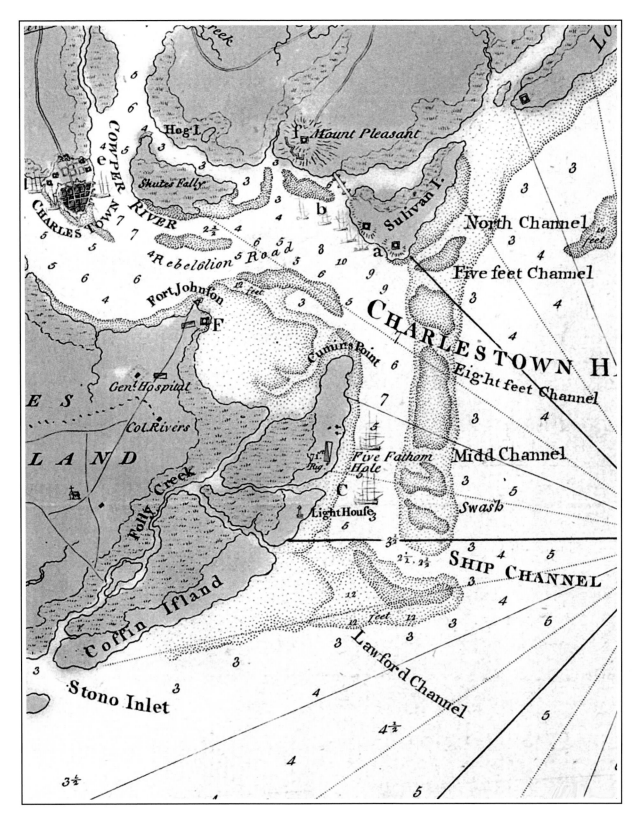

FIGURE 122. Detail of a 1780 map of Charleston Harbor made by the British military. Note how the large ebb-tidal delta off the entrance to the harbor is deflected far to the south. Compare this map with the image of the ebb-tidal delta of North Inlet shown in Figure 101. Map © National Maritime Museum, Greenwich, London.

shorelines retreated "at approximately four to six feet per year."

- By the second half of the 20th century, beach erosion became a continuing theme in the Folly Beach area and a number of beach cottages were lost to erosion. Therefore, a series of groins was constructed along the beach of Folly Island in an effort to combat the erosion. The first groin construction began in the 1940s. These were spaced 1,000 feet apart and by the 1950s, a second series of groins was added between the original groins, resulting in a groin every 500 feet. These remained at least partially functional until hurricane *Hugo* struck the S.C. coast in 1989 (Lennon, 2000). The main problem with the groins at Folly Beach, besides the obvious problem of limited sand supply, is that they are oriented directly into the dominant wave approach direction, thus their function of trapping sand moving laterally along shore is severely impaired. The erosion on Folly as a result of *Hugo* and the utility of groins were discussed further in Section I. According to Dr. Tim Kana, one reason for the "perception" of failure of the groins is that "if you want to stabilize a beach that averages 300 feet wide at low tide, the groins need to be about 400 feet long. Folly Beach's groins average about 250 feet long." Furthermore, "they weren't properly maintained and leaked sand through holes in the timbers, further reducing their effectiveness."

When our group did an erosion survey of Kiawah Island in 1974, we were curious why that island, which is located only a few miles south of Folly Island, suddenly began to accrete significantly on its northern end in the early 1900s (see Figure 123). This curiosity led us to examine the historical impact to the shoreline of the jetty construction at Charleston Harbor. Our conclusion was that a significant part of the mass of sand that first was deposited on Folly Island and eventually eroded

had made its way to Kiawah. We published a paper noting the impacts of the jetties described in the previous section, but the U.S. Army Corps of Engineers office in Charleston was reluctant to accept that premise, noting that the beach to the southwest had accreted after the jetties were built. That observation was true, of course, but ultimately, with no sand passing around the jetties, the shoreline began to retreat.

In any event, Hayes made many trips to Folly Beach at the request of the mayor, Richard Beck, to give lectures, *pro bono*, on this and other matters related to the erosion at Folly Beach. Long after we had departed that scene, the powers that be agreed with at least some of our original conclusions about the jetties, and programs to nourish the beach were adopted. These nourishment projects up through 2000, which were summarized by Lennon (2000) and T. W. Kana (pers. comm.), are briefly outlined below:

- In 1993, the U.S. Army Corps of Engineers completed the first stage of a 50-year project. This project was planned with the intention of nourishing every eight years. In the first stage in 1993, twelve to fifteen million dollars were spent on Folly Island's beach nourishment projects. Approximately 3 million cubic yards of sand were pumped from the backside of the island and deposited over a length of 5.3 miles, from the south tip to the middle part of the shoreline.

- By 1995, approximately two thirds of the added sand had eroded according to Lennon. However, Kana disputes this claim, noting that only 20% of the sand was lost during the first two years. However, during this period, the county park, at the southern end of the island, suffered a significant loss of sand.

- An emergency renourishment of 90,000 cubic yards of sand was implemented in April 1998 at the southern end of Folly Island due to the advanced erosional state of the county park.

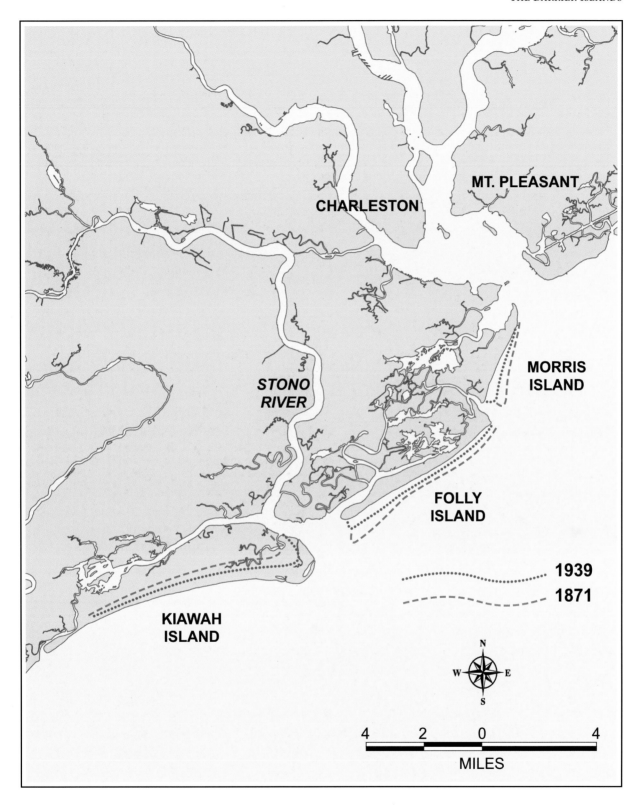

FIGURE 123. Historical shoreline changes, Morris Island to Kiawah Island. Morris and Folly Islands eroded hundreds of feet and Kiawah Island prograded more than 3,000 feet during this time period. This diagram was created in 1974, but the trend in shoreline changes shown continued until the 1980s. According to a recent report by DuMars (2007), "a reduced rate of retreat since 1933 and recent accretion suggests that Morris Island's shoreline has reached equilibrium." Since 1993, several beach nourishment projects have been carried out to combat the beach erosion on Folly Island.

There was some litigation involved with this.

- Less than one year later, additional nourishment was required because sand from the 1998 nourishment was gone and the park's acreage was continuing to erode. Only 130 of the original 400 parking spaces at the park were left (as of December, 2001). To remedy the situation, 49,000 cubic yards of sand were dredged from the Folly River, enough for a 150-foot wide beach 1,000 feet long (Lennon, 2000).

- Another 2 million cubic yards of sand were pumped onto Folly Beach in December 2005.

According to W. J. Sexton, who is in the business of finding sand for such nourishment activities, the sand for the late 2005 project came from offshore in the area off Lighthouse Inlet where the original ebb-tidal delta of Charleston Harbor was located. When we visited the north end of Folly Island in January 2007, the beach was quite wide.

Morris Island, located just to the north of Folly Island, has undergone even more extensive erosion than Folly Island since the jetties were constructed. This spectacular erosion, which is illustrated in Figure 123, has caused Morris Island to be offset even further landward than Folly Island (Figures 121 and 123). However, a recent study by DuMars (2007) indicates that the beach on Morris Island is relatively stable at the present time.

An oblique aerial photograph of the lighthouse area on the south end of Morris Island is given in Figure 124. The lighthouse, which now sits several hundred yards offshore of the island, is of particular historic interest. As noted on the web site for Save the Light, Inc., in 1673, only three years after Charles Town, the original European settlement in the Charleston area, was settled, a navigation aid consisting of a fire that burned at night was established on the island. A lighthouse 42 feet tall was built in 1767, but it was replaced by a new lighthouse in 1838 that was 102 feet tall. The lighthouse built in 1838 was destroyed by the

Rebels during the Second Civil War (in 1862) so the Yankees couldn't use it. In 1876, a new lighthouse was constructed 1,200 feet back from the high-tide line, but by 1938, it was being lapped at by waves at high tide. Retreating further from shore as the island continued to erode, the 1876 lighthouse was finally decommissioned in 1962 and replaced by the new Sullivans Island Light, which you may have seen if you visited Sullivans Island. In 1962, the Morris Island light was sold to private buyers, but at the present time, it is "owned" by the South Carolina Department of Natural Resources (see the web site www.savethelight.org for more details).

Amazingly, the lighthouse did not budge during hurricane *Hugo* (1989). Well, okay, maybe it did tilt a little bit. The lighthouse is still there in 2007, because its foundation extends 35 feet below sea level (Pilkey and Fraser, 2003). The distance to the lighthouse from shore can be expected to change some in the future, depending upon the vagaries of the ebb-tidal delta of the small tidal inlet just to the south of the lighthouse (Lighthouse Inlet), which is shown in Figure 124.

At times during the Second Civil War, as many as twenty thousand Union troops occupied a fort on Morris Island called Fort Wagner. This fort was the site of a fierce battle in which African-American troops participated on the Union side. This battle was the "final desperate battle depicted in the movie *Glory*" (Pilkey and Fraser, 2003).

Places to Visit

To get to Folly Beach you go south from Charleston on US 17, cross the Ashley River and almost immediately take a left on state road 700, which joins up with state road 171, and follow that road all the way to the island. At mile 0.7, you cross over the Intracoastal Waterway at Wapoo Cut, an artificial shortening of the waterway through a Pleistocene upland that connects the Ashley River with the Stono River, two coastal plain rivers. The Stono River has very little fresh water input, being

FIGURE 124. Lighthouse Inlet, which separates Folly and Morris Islands, at low tide in October 1978. The main ebb channel of the inlet cuts obliquely across the center portion of this infrared image. Note waves breaking around the seaward extent of the ebb-tidal delta. A large swash bar complex had developed on the northern channel margin linear bar of the ebb-tidal delta (compare photograph with model in Figure 37). A well-defined marginal flood channel (arrow) was situated between the swash bar complex and the beach. The Morris Island Lighthouse, which was constructed 1,200 feet landward of the high-tide line in 1876, was located far off the severely eroding beach when this photograph was taken. According to DuMars (2007), this inlet migrated to the southwest at the rate of around 5 feet per year between 1900 and 2006, and is "expected to continue" at that rate into the future.

primarily a tidal river at this time in its history. On the way to Folly Beach, you will pass through the highly populated James Island, one of the triangular shaped "sea islands" underlain by Pleistocene shoreline deposits that populate the southern half of the state's coastal area.

As soon as you cross the bridge over Wapoo Cut, that is where you turn right to go to Kiawah and Seabrook Islands. Continuing on toward Folly Beach, at 6.1 miles you encounter a wide marsh area that marks the seaward edge of James Island. There are some scattered Pleistocene highs in this backbarrier area. Then you cross a wide zone of

dwarfed smooth cordgrass (*Spartina alterniflora*). The patches of mud flats along the road in this marsh area are present where high-tide wrack lines have killed the marsh grass.

At 8.6 miles, you are on the barrier island itself. Drive straight to the traffic light and turn right should your destination be the **Folly Beach County Park**. Driving along the road toward the park, which is just behind the front row of beach cottages, there are nine public beach access points where you can walk between the beachfront cottages to the beach, but there is limited parking space. This is a visitor-friendly island, thanks to the efforts of our friend and former mayor of the island, Dr. Richard Beck. Most of those houses on the beachfront you are driving by are post-*Hugo*. There are numerous vacation rental houses along this street.

Folly Beach County Park is an appealing beach system on the end of the recurved spit on the southern end of Folly Island. At one time, there was talk about developing this spit for beach houses, which we recommended against because of the narrow width of the spit and the erosional history of the island. This is a very nice, clean beach with an interesting walk down to the end where you can typically view intertidal bars and other aspects of the beach morphology. However, there is a strong longshore current along this beach, so swim with caution. The birding is good there as well, with shorebirds, wading birds, pelicans, and cormorants

typically being present in abundance.

The road onto the island leads directly to the Holiday Inn, which was remodeled after hurricane *Hugo*. For years before the hurricane, we used to bring groups in our training courses for dinner at the Atlantic House Restaurant, located in an old wooden building built up on wooden pilings out over the water. It was located about a block to the south of where the Holiday Inn is now. We were always there at high tide, and it was a unique experience to feel the big waves roll by under the building on their way to break on the beach. This restaurant is shown in the oblique aerial photograph in Figure 125A. The post-*Hugo* picture in Figure 125B shows there was not a trace left of the restaurant after the storm. Too bad, it was a perfect spot to end the day.

Another place you shouldn't miss on Folly

FIGURE 125. Folly Beach before and after hurricane *Hugo* (1989). (A) Photograph taken circa 1980. Arrow points to the Atlantic House Restaurant. (B) Arrow points to former location of the Atlantic House Restaurant. Photograph taken 11 October 1989, 19 days after the passage of hurricane *Hugo*.

Island is a visit to the tidal inlet on the north end of the island and its associated Morris Island Lighthouse (Figure 124). From the intersection where you turned to go south to the park, drive northeast along the main road. You will notice more public beach accesses as you drive north along this road.

At 1.8 miles north, you will reach an area the locals call either "surf beach" or "the washout," which is protected by riprap at this time. This part of the island was completely washed through during *Hugo*, consequently, there are no beachfront houses in this section. At this site, there is space for parallel parking along the road, as well as several boardwalks over the riprap to the beach.

As shown by the photograph in Figure 126, the beach was very wide and flat and the groins were partially buried by sand when we were there in January 2007. At this part of the island, you can also observe the efforts to keep sand on the beach (and protect the road) with two rows of snow fences and planted sea oats *(Uniola)*. When we walked out onto the beach, we observed a lot of shell material and heavy minerals in this recently nourished beach sand.

At 2.2 miles further along, more beach cottages are located seaward of the road, some of which apparently did survive *Hugo*. At 3.2 miles, the road ends at a parking area where a gate prohibits further driving to the north. However, from this parking

FIGURE 126. Beach view looking southwest on the northeast sector of Folly Beach. The line of dune sand trapped by the snow fence is topped by sea oats *(Uniola)*. Photograph taken on 13 January 2007.

area, we highly recommend that you walk the few hundred yards to where you come to the side of a tidal inlet that empties into the open ocean. This also gives you the opportunity for a look out to the Morris Island Lighthouse, which, as noted above, was last on the beach itself back in 1939. Also, if you go at low tide, you can see the exposed sand bars associated with the ebb-tidal delta of this small tidal inlet (see oblique aerial photographs of the lighthouse and inlet in Figure 124). This is also a great place to see shorebirds and other avifauna if you visit at low tide.

KIAWAH AND SEABROOK ISLANDS

Although these islands go by different names, they are actually both part of a single large prograding barrier island complex, separated only by a small tidal inlet (Captain Sams Inlet), the outlet for a moderate-sized tidal creek that drains the marsh system behind Kiawah Island. By examining the image of the area shown in Figure 35, you will see several significant contrasting features with regard to the morphology and human development of these islands:

- The parallel beach ridges on Kiawah Island signifying the progradational character of the island, which also accentuate its pronounced "drumstick" shape (see also the oblique aerial photographs in Figures 23, 28, and 34).
- The two large tidal inlets on the northeast and southwest ends of the barrier island complex and the smaller Captain Sams Inlet (Figure 34) that separates the two islands.
- The contrasting backbarrier regions of the two islands, with the Kiawah Island backbarrier being composed entirely of salt marsh, tidal channels, and tidal flats and almost half of the backbarrier of Seabrook Island being attached to a remnant Pleistocene upland.
- The evidence of human development on both islands, including several golf courses.

- Massive sand accretion on the north end of Kiawah and the large bare sand shoal/island (Deveaux Bank) on the ebb-tidal delta of North Edisto Inlet (southwest end of Seabrook; Figures 35 and 38A).

KIAWAH ISLAND

History

The history of Kiawah Island, which is about 14 miles long and a mile and half across at its widest point, is fascinating in all its many aspects. The island began to take shape about 4,500 years ago, when sea level stopped rising after the last ice age and became relatively stabilized at near its present level. For the most part, the shoreline of the island has been building out in a seaward direction ever since. Unraveling this history was one of the projects our group initiated in 1974, when we began a long sequence of studies on Kiawah. One of the end products of that early work was a dissertation project by a graduate student of Hayes', Tom Moslow, who was a champion of the *transgressive/ regressive interfluve hypothesis* (Figure 33) for the origin of barrier islands discussed above (named as such by us, not by him; Moslow, 1980)

The island is named after one of the many small Indian tribes that occupied the land now known as South Carolina before the white man started arriving in force around 1670. The Kiawah tribe is now history, as are most of the other lesser tribes, such as the Congarees, Sewees, and so forth.

During the First Civil War, the British captured the island fairly early in the war and held it until the end. A few years later, General Arnoldus Vanderhorst, a hero of the First Civil War and twice mayor of Charleston, built his home on the eastern half of the island. Eventually, the entire island passed from one generation of Vanderhorsts to the next, who successfully farmed sea-island cotton until the Second Civil War. During that second war, the island was captured by the Yankees, and, in fact, we are told that a black Union regiment was encamped on the island for some time (presumably the same one that fought the famous battle at Fort Wagner on Morris Island). When we did our first survey of the island in 1974, the old Vanderhorst mansion was in rather poor shape, but the walls on the second floor were covered by graffiti written in gorgeous large script by the Union soldiers – "WE DON'T MUCH LIKE YOU GENERAL BEAUREGARD" (the Confederate general in charge of Charleston at that time) and things like that (edited version given here).

To finish the story, in 1950, the Vanderhorst estate sold the island to C.C. Royal, a lumberman from Aiken, South Carolina, for $125,000. Over the next 24 years, Royal logged pine trees, built the causeway to Kiawah, and sold a few beachfront lots. Also, several roomy and elegant houses were built on beachfront lots near the middle of the island. In 1974, Royal sold the island to the Kuwaiti Investment Corporation for $17 million. At that point, the second largest barrier island in South Carolina was owned by investors from a foreign country. As soon as this investment was made, the development of the island went into high gear. Finally, in 1988, Kiawah Resort Associates purchased Kiawah Island (the management role and that part of the island not already sold to new residents and related businesses) from the Kuwaitis for $105 million, and in 1988, the Town of Kiawah Island was incorporated.

Geomorphology and Erosional/Depositional History

After the Kuwaiti investors purchased the island in 1974, Charleston County applied rather strict development regulations. These regulations required an environmental impact statement, which incorporated the development of construction set-back lines for all the beachfront property. A fellow faculty member at USC-East, Dr. John Dean, and his partners formed a new consulting company and

landed the environmental project, and our group was awarded the job to develop the construction set-back lines. We also did a basic geomorphological study of the island, with the map in Figure 127 of the island's geomorphic provinces being one of the products of the study.

The set-back lines we drew were based primarily on two criteria: 1) a detailed analysis of the shoreline changes on the island, starting with historic charts dating back to 1661 (six maps and sketches and five coastal charts) and five sets of vertical photographs taken between 1938 and 1973 (data summarized by Stephen et al., 1975); and 2) historical beach erosion rates during hurricanes elsewhere, relying heavily on Hayes' work on hurricane *Carla* (1961) (Hayes, 1965; 1967). The goal of the set-back concept was to draw a line beyond which the beach was not expected to erode within the next 50 years. For the record, it is now 33 years since those lines were drawn, and significant erosion has not occurred on the island where the original setback was adhered to, and to this date, no structures have been lost due to erosion. There is presently concern about the erosion of the beach along the Ocean Course (golf course; T. Kana, pers. comm.), but the golf course was built seaward of the original set-back line. If the eye of hurricane *Hugo* (1989) had gone to the south of the island, no doubt severe wind damage would have occurred, but it is doubtful that any beachfront lots would have been lost, as was the case on most of the Isle of Palms during hurricane *Hugo*. Erosion of the Kiawah beach was minor during *Hugo*.

The data we analyzed on shoreline changes revealed the following (Hayes et al., 1975):

- The central portion of the island had shown the least change of any part of the island, and it had been slowly prograding throughout the study period. This was no big surprise, considering the long-term progradational history of the island. Between 1867 and 1973, the west-central shoreline prograded about 600 yards and the

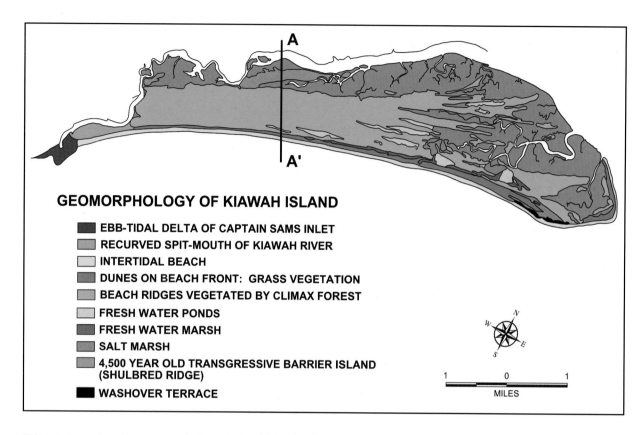

FIGURE 127. Map of the geomorphology of Kiawah Island, using a 1971 aerial photograph as a base. The line labeled A-A' gives the approximate location of the stratigraphic section shown in Figure 132.

east-central shoreline advanced more than 1,800 yards. Data for the period 1934-1993, summarized by Kana and Gaudiano (2001), showed that the west-central area eroded only slightly (about 2 cubic yards per linear yard of beach/year) and that the east-central shoreline continued to accrete (about 12 cubic yards per linear yard of beach/year).

- The southwestern part of the island is a recurved spit, which is pictured in Figure 23. The historical changes of the spit, given in Figure 128, show that about every 40-50 years, the neck of the spit is cut through, most likely during a hurricane. After the cut, the spit once again begins to grow and the adjacent inlet, Captain Sams Inlet, migrates to the southwest at a rate of around 100-200 feet per year.

- The northeastern, or updrift, end of Kiawah Island had undergone many changes in the

past 300 years. In 1661, a large waterway incised the northeastern portion of the island. Between 1661 and 1854, the waterway filled in with sediment. Beginning in the late 1880s and continuing at a rapid rate until the 1920s, the northeastern shoreline underwent rapid accretion. This trend continued at a slower rate until the late 1930s, adding a total of over 3,000 feet to the shoreline in the form of a triangular foreland. As noted earlier, no doubt a good deal of this sand was eroded from Folly Island, which had initially received it as a result of the huge ebb-tidal delta off Charleston Harbor coming ashore (after the jetties were built in 1889). This foreland eroded relatively slowly between the 1930s and the time we did our survey in 1974. Since 1974, however, a series of large swash bars have welded to the northeast end of Kiawah Island (see Figures 56A and 129),

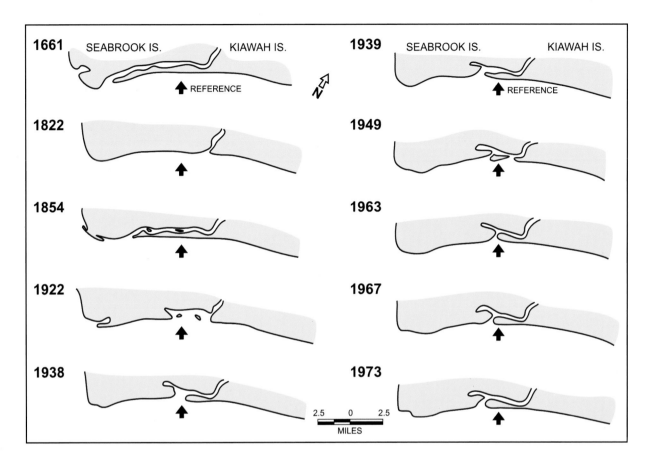

FIGURE 128. Historic changes Captain Sams Inlet between 1661 and 1973, based on a combination of historical charts and aerial photographs. These maps and photos show three periods of breaching of the spit: 1822, 1922, and 1949. Breaches occurred at numerous other times, of course, at intervals we estimate to be about every 40-50 years. From Hayes et al. (1976).

resulting in an amazing amount of accretion on that part of the island. As noted by Tim Kana, "*in barely ten years, a 3-mile-long barrier island/lagoon system formed by way of shoal bypassing at Stono Inlet. An estimated 5 million cubic yards were added by natural processes at Kiawah's eastern end. As the bypassing event progressed, it enclosed a 200-acre lagoon, added 150 acres of beach/dune/washover habitat, and left a flushing channel in front of the Ocean Course Club House.*" Concern about the outlet channel moving too close to the dunes by the golf course called for a solution to the problem, which was recommended and carried out by cutting a new, shorter outlet channel straight across the massive welded bar complex and

moving some of the huge volume of sand available closer to shore (under the direction of Coastal Science and Engineering; Figure 130). It will be interesting to learn the final results of this project.

For the record, the set-back line constructed by us in 1974 was drawn a modest distance from the high-tide line in the middle of the island, a few hundred feet, but the two extreme ends, adjacent to the tidal inlets, were given a wide berth. In fact, the line for the recurved spit on the south end of the island was drawn at the neck of the spit, and the developers wisely chose to not develop the spit at all. The spit eventually was deeded to Charleston County to make a park for public use (Beachwalker

STONO INLET

FIGURE 129. Oblique view of north end of Kiawah Island, showing a large swash bar welding to the beach. This is the main mechanism by which such prograding barrier islands attain their drumstick configuration (see also Figures 28, 29, and 56). Infrared photograph taken at low tide on 29 May 1980.

Park). All of the land in the photograph in Figure 130 is located seaward of the original set-back line, so the only development seaward of the line is the golf course. However, that is not to say that the golf course should not have been developed where it is. We do not think there will be a significant long-term erosion problem at this site, as long as: 1) the sand keeps coming down the shore at the rate it has been in recent years, and 2) sea level does not rise abruptly.

The Beach

Following is a description of the beach of Kiawah Island based mostly on a write-up we did in 1976. We have visited the island many times since

then, mostly while conducting geological training seminars, and these descriptions are consistent with the present condition of the beaches. To our knowledge, there are no seawalls or revetments on the outer beach.

In 1974, the beach zone in the middle of the island consisted of three components: 1) a series of parallel beach ridges located immediately behind the foredune area ranging in age from 50 to a few hundred years. Further landward on the landward side of the island, the oldest beach ridges approach 4,000 years in age. In 1974, almost all of these beach ridges were host to a climax maritime forest dominated by live oaks (*Quercus virginiana*), palmetto (*Sabal palmetto*), and loblolly pine (*Pinus taeda*) (Sharitz, 1975). Near the beach, this

204

FIGURE 130. The Ocean Course on the north end of Kiawah Island. This photograph by Tim Kana was taken at low tide on 13 July 2006, after the completion of an erosion remediation effort by his company, Coastal Science and Engineering, LLC. A few months earlier, a very large swash bar complex similar to the one shown in Figure 129 approached the shore. As a result, the flow from a tidal creek draining a marsh system was diverted to the south (arrow A) such that it hugged the beach for several hundred yards, allowing the front dune line to erode. The remediation involved cutting a new channel across the welded swash bars (arrow B) so that the channel would no longer impact the beach, as well as moving part of the mass of sand toward the beach. Eventually, most of the sand in the upper right hand part of the photograph will weld to the beach.

vegetation shows considerable salt pruning. In many places along the beach at the present time, these trees have been at least partly replaced for home sites. In many areas, our set-back line was drawn at the edge of the climax maritime forest, so there usually are no houses seaward of this line (there are a few exceptions). The next zone is a 200-300 feet wide area composed of several dune lines. These dunes are dominated by grass vegetation on the seaward portion and contain some scattered wax myrtle *(Myrica cerifera)* on the more landward part. Commonly, there are at least a couple of dune lines in the pure grass area. The land where the wax myrtle is growing was at least 15 years old in 1974 (P. Hosier, pers. comm.). The front foredune ridge usually rises 5-10 feet above the swash line of high spring tides. The vegetation in these dunes is dominated by sea oats (*Uniola paniculata*), sea elder (*Iva imbricata)*, and *Croton sp.,* with the wax myrtle, where present, growing in swales between the higher dunes. The intertidal beach zone is an exceptionally wide, fine-grained flat beach that occasionally contains low intertidal bar and trough systems (Figure 53). The intertidal zone averages approximately 250-300 feet wide at low spring tides. A series of topographic profiles and an oblique aerial photograph of the beach in 1974 are given in Figure 131.

As illustrated earlier, the extreme northeastern end of the island has changed dramatically over time. Figure 129 shows one of the large intertidal bars that have been common in this area. As recently as February 2007, very large swash bars continued to migrate toward the backbeach. If you are on the island as a guest, you can reach this area by walking northeast along the beach from the Ocean Golf Course, the site of at least two Ryder Cups and the 68th Senior PGA Championship on 22-27 May 2007. This is a spectacular area to visit, if only to envision the huge amount of sand that has come ashore in the past couple of decades. Such a visit will also give you a chance to see how Kana's beach-erosion remediation project is coming. It is also a

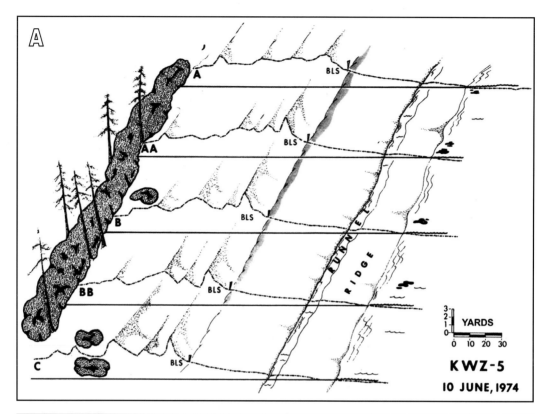

FIGURE 131. Beach morphology of the middle of Kiawah Island, a typical prograding, mixed-energy barrier island. (A) Three-dimensional view of the beach at low tide, illustrating a low intertidal bar and trough and prograding foredune ridges (on 10 June 1974). Diagram is based on five beach profiles spaced at 150-yard intervals and plotted at a 5:1 vertical exaggeration. BLS = stake near the high-tide line. (B) Oblique aerial view of the beach at low tide in July 1974, looking northeast. This area is approximately 1.5 miles north of the one mapped in A.

206

complex natural area with a unique combination of coastal habitats.

The **recurved spit** at the southwest end of the island is composed of recurved vegetated dunes and a wide beach zone. Captain Sams Inlet is located several hundred yards down the beach from the spit's neck (see oblique aerial photograph in Figures 23). The dune ridges are distinctly linear and curve in toward the tidal channel behind the spit. These ridges are continuous for several hundred yards and attain a height of around 10 feet above the last high-tide line. The vegetation in the dunes is similar to that described for the dunes in the middle of the island. The intertidal beach is wide and flat, averaging over 300 feet wide at low spring tide. It is common for one or two low-amplitude intertidal bars to be present on this beach.

Tidal Inlets

The erosional and depositional history of Kiawah Island is closely related to changes in the morphology and processes associated with its neighboring tidal inlets, Stono on the northeastern end and Captain Sams on the southwestern end. Considering the Kiawah/Seabrook complex as a whole, the huge inlet at North Edisto Inlet to the southeast of Seabrook also comes into play. To compare the volumes of sand involved in these systems, we calculated that the two large ebb-tidal deltas of the inlets at the ends of the islands, Stono and North Edisto, contains approximately 200 million cubic yards of sand, 78% as much sand as is contained within the two barrier islands themselves.

Waves approaching the adjacent beaches of the two barrier islands are strongly influenced by these huge masses of sand. Wave refraction around the two ebb-tidal deltas has played a role in shaping the beaches of both of the islands. The ebb-tidal deltas are also important as large storage areas for sand, as is clearly obvious based on the rapid accretion at the northeastern end of Kiawah Island.

Salt Marsh and Tidal Channels

The entire northern boundary of Kiawah Island is surrounded by extensive salt marshes and tidal creeks. Salt-marsh grasses are composed mainly of smooth cordgrass (*Spartina alterniflora*) in the intertidal zone and saltmeadow cordgrass (*Spartina patens*) and bushy seaside tansy (*Borrichia frutescens*) in the supratidal areas (see Figure 71 for a photograph of the marsh on the landward side of Kiawah Island). There are areas of somewhat extensive tidal flats in the north-central portion of the island and oyster mounds (*Crassostria virginnica*) are quite abundant in some areas. Although the major tidal creeks migrate rather slowly, there are numerous examples of steep cut banks where the creeks have eroded into the old beach ridges (Figure 76A). Tidal currents are usually quite strong in the major tidal channels, attaining velocities of one or two knots, and the bottoms of the creeks are generally floored by sand.

Stratigraphy

The geomorphology of this prograding barrier island is illustrated in Figure 127. The most landward component of the island is Shulbred Ridge, the primary landward-migrating element of the island, which was postulated by Moslow (1980) to have formed around 4,500 B.P. The sediments of Shulbred Ridge are medium- to coarse-grained, poorly sorted sand that is reddish in color. The bulk of the rest of the island is composed of beach ridges that have prograded seaward in the past 4,500 years. Analysis of 180 samples from the prograded beach ridges showed them to be homogeneous fine to very fine-grained sand throughout (mean size = 0.15-0.2 mm; Hayes et al., 1976). The striking difference in the character of the sand in Shulbred Ridge versus the other beach ridges influenced Hayes to postulate in 1974 that Shulbred Ridge was a Pleistocene deposit. Age dating later proved that idea to be wrong. Shulbred Ridge is Holocene in age, a discovery that did lead us to understand the significance of the original landward-migrating element (i.e., Shulbred Ridge) in the formational history of these barrier islands, as is demonstrated in Figure 33.

The stratigraphy of the island was studied in detail by Moslow (1980). A generalized stratigraphic section through the middle of the island is given in Figure 132, which demonstrates three significant components to its stratigraphic evolution:

1) A lag deposit of coarse shell material usually < 2 feet thick deposited on an erosion surface (**transgressive surface of erosion**) created as the original barrier island migrated landward. This deposit presently occurs everywhere beneath prograded shoreface sand. The shoreface is the seaward face of the barrier island

between mean low tide and typical storm "wave base" (about 30-foot depths).

2) Deposition of the primary, landward-migrating barrier island (Shulbred Ridge) at the landward side of the island around 4,500 B.P. (at the initiation of the present stillstand of sea level).

3) After initiation of relative stillstand conditions, seaward progradation of a 20-30 feet thick, 1.5 mile wide sand body. Moslow used woody material collected from cores across the island to determine the age of each of the beach ridges.

While the process of Holocene beach-ridge accretion was taking place, a minor inlet (Captain Sams) formed between Kiawah and Seabrook Islands. In addition, migrating tidal creeks began eroding the oldest beach ridges (see Figure 76A).

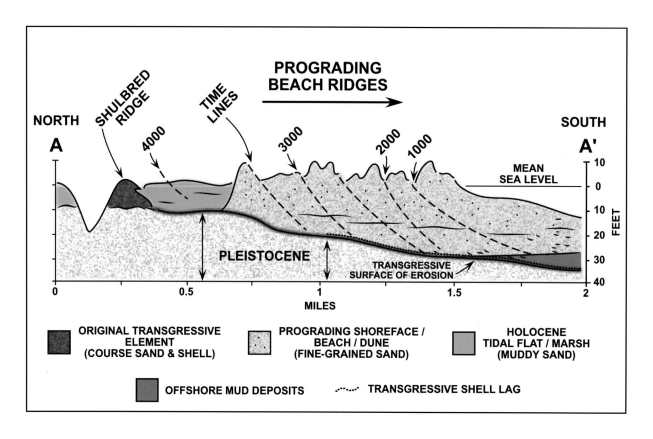

FIGURE 132. Generalized stratigraphy of Kiawah Island. Modified after Moslow (1980). This cross-section is located (approximately) on Figure 127.

Places to Visit

To get on the beach on Kiawah, short of going there by boat, you have two choices: 1) become a guest on the island; or 2) visit **Beachwalker Park**. There are also five golf courses that allow off-island users (for a fee, of course). Whichever method you chose to get on the beach, we recommend highly that you acquire a bicycle (there are rentals on the island and at times at the park) and ride the full 14 miles of the beach. This is one of the most beautiful beaches in the eastern United States. Again, low tide is the best time to go – easier riding because the sand at the high-tide line tends to be softer, you can see the entire expanse of the intertidal zone, and the shorebirds are out in force.

Beachwalker Park can be reached by driving onto the island and turning right just before reaching a service station located immediately outside of the main entrance gate. This quiet setting offers excellent access to this splendid beach, and swimming is available with seasonal lifeguards.

Hours for the park are complicated:

- Mar - Sat - Sun, 10 am-5 pm
- Apr - Sat - Sun, 10 am-6 pm
- May - Labor Day - Daily, 9 am-7 pm
- Sep - Daily, 10 am-6 pm,
- Oct - Sat - Sun, 10 am-6 pm
- Nov-Feb - **Closed**

A boardwalk offers the possibility for a pleasant walk across the foredunes. Watch in the myrtle and young pine trees for exotic birds, such as the **painted bunting,** which nest in the area in the spring. A photograph of this spectacularly colored bird is shown in Figure 133. Additional amenities in the park include restrooms, showers, snacks, picnic areas with grills, and rental facilities for umbrellas, chairs, and bicycles (at times).

Bicycle or no bicycle, by all means take a trip down to the entrance to Captain Sams Inlet, where you can observe the exposed tidal flats of the inlet margin and walk around the back side of the

FIGURE 133. The painted bunting. Photograph courtesy of the Friends of Hunting Island.

recurved spit in order to visit the tidal channel and associated salt marsh. On the tour down to the inlet, you will be able to observe the different aspects of the beach on the spit sketched in Figure 134, which include the following:

- A very well-developed foredune ridge, which, at times, may have an erosional scarp, covered with abundant dune grasses, including sea oats (*Uniola*) on the top of the dunes and sea elder (*Iva imbricatta*), and *Croton sp.,* among a few others, around the base.

- Between the base of the dunes and the high-tide line, you usually see abundant burrows of the ghost crab (*Ocypode quadrata*). According to the Wikipedia free encyclopedia, *"These crabs are called ghosts because of their ability to disappear from sight almost instantly, scuttling at speeds up to 10 miles per hour, while making sharp directional changes. These creatures have two black eyes, with sharp 360° vision."* As you walk along the high-tide line, you will see numerous mounds of sand excavated from the

ghost crabs' J-shaped burrows, which tunnel down into the ground at a 45 degree angle as much as 3 feet or so. If you are patient, you will probably see a few scurrying about. See photographs in Figure 135.

- Moving down the beach profile from the high-tide line, you first cross the high beachface, which commonly contains antidune traces (see Figures 50 and 51).

- Lower down, near the mid-tide line, it is common to encounter an intertidal trough, which usually contains sand ripples (Figure 46A). At times, these troughs still hold water at low tide and you can see the current draining the trough.

- Finally, you will walk up onto one, or sometimes two, intertidal bars, which typically are low-amplitude bars on this beach.

- The lower half of the intertidal zone usually contains abundant burrows of the ghost shrimp (*Calianassa major*), the most we have seen on any beaches in the state (see examples of burrow outlets in Figure 136). As the

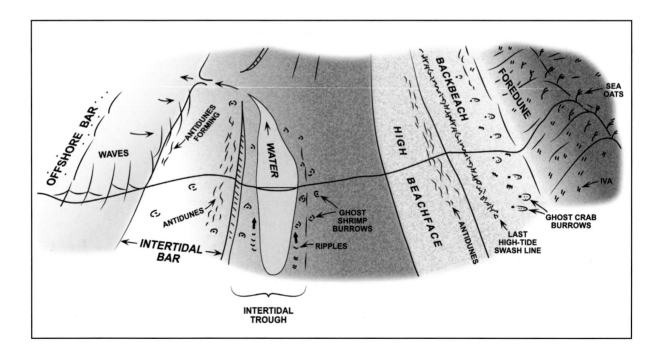

FIGURE 134. Sketch showing the different aspects of the beach exposed at low tide on the recurved spit on the south end of Kiawah Island.

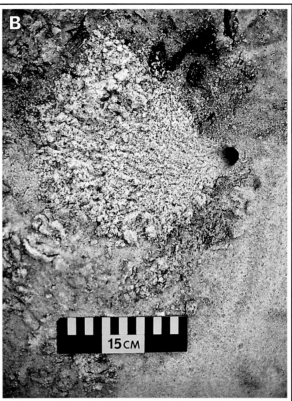

FIGURE 135. The ghost crab (*Ocypode quadrata*). (A) A rare view of one on the surface in the daylight. (B) Sand castings outside one of their burrows.

photograph in Figure 137B shows, the burrow outlets are usually surrounded by numerous chocolate-jimmie-like pellets composed of organic-rich mud. These are the fecal pellets of this rapidly burrowing shrimp (pictured in Figure 137A), which is almost impossible to catch. The shrimp use these pellets to line their corncob-like burrows in order to strengthen them. A picture of some of these burrow tubes exposed in a Pleistocene barrier island at the Florida/Georgia border, in the lower intertidal part of that beach, attests to the ruggedness of the burrows (see Figure 137C). These types of burrows are called *Ophiomorphia* where they are found fossilized in ancient sedimentary rocks. For example, they are abundant in the Cretaceous sandstones of western Canada, which are as much as 100 million years old.

- If you don't mind getting wet, you can continue

further out into the water until you walk up onto another bar, the nearshore bar, where you may be able to feel the bedforms (ripples and megaripples) under your feet in the trough before you walk up onto the bar.

SEABROOK ISLAND

Discussion

This island is a gated community, which is a very pleasant place to stay if you are so inclined. The geomorphology, stratigraphy, and so on of this island are similar to Kiawah. It has had some erosion problems, but no set-back line approach was ever attempted. Like Kiawah, the development of this island for residences began in the early 1970s. In 1978, our group was contacted by the developers regarding their shoreline erosion problems, of

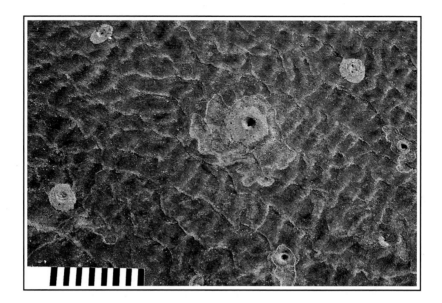

FIGURE 136. Outlets of ghost shrimp burrows on the lower intertidal beach on Kiawah Island. Scale is in centimeters.

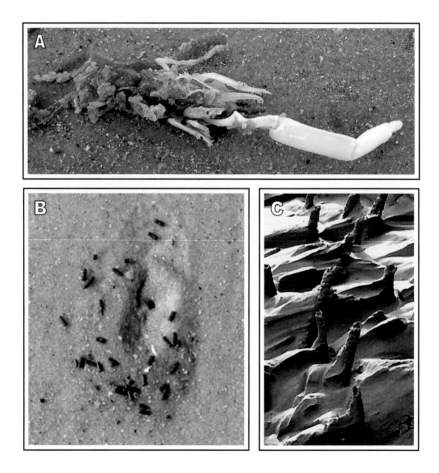

FIGURE 137. The ghost shrimp (*Calianassa major*). (A) Live ghost shrimp with its large "digging claw" exposed. (B) Fecal pellets accumulated around a burrow outlet. Photographs A and B courtesy of Bill Frank, Jacksonville, Florida. (C) Corncob-like burrows exposed by erosion that were fortified with fecal pellets like the ones shown in B. These dense and spectacularly displayed burrow remnants were originally formed in the lower intertidal zone of a Pleistocene barrier island that is now exposed along the banks of the St. Marys River on the Georgia/Florida border.

which there were three:

1) The original Beach Club and marketing center was built on the side of the main channel of the North Edisto Inlet, which attains depths of over 70 feet in the inlet throat. As shown by the photograph in Figure 138A, the beach was erosional in the fall of 1974. We were not involved with the solution to this problem, which was to build a revetment composed of large riprap along the side of the channel (Figure 138B). The building in the 1974 picture was still there in February, 2007.

2) As shown by the oblique aerial photographs in Figure 139, an eroding arc in the beach in the middle of the island was threatening the golf course, as well as houses nearby. This was erosion caused by waves approaching the shore from the open ocean.

3) Captain Sams Inlet was migrating into the island, with a loss of many acres per year of highly developable land as a result of erosion. In 1982, we advised the developers that the inlet would be at the golf course in 14 years.

A review of historic charts, maps, and aerial photos showed that the outlet of Captain Sams Inlet (Figures 34 and 128) had migrated south and eroded large parts of Seabrook Island several times. Hayes also noticed that the neck of the spit was last cut a fairly short time before the 1949 vertical aerial photograph was taken. The cut may have happened during a hurricane a year or two earlier, but we have not been able to determine which one, if indeed a hurricane was responsible. It is possible that the migrating meander bend of the large tidal creek behind the spit eroded into the foredunes to such an extent that waves generated by a relatively minor storm could have eroded through the spit during an unusually high spring tide. We think the odds favor a hurricane, we just don't have enough data to say which one. No major hurricane crossed the South Carolina coast during that time, but the

eyes of a couple did cross the Georgia coast, which means there should have been a significant storm surge along this part of the South Carolina coast.

Based on studies of the aerial photos taken soon after the new inlet had cut through at the neck of the spit in the late 1940s, Hayes observed that the shoreline of Seabrook Island built out dramatically within a few years afterwards. At the time of that 1940s cut, the inlet throat was near the spot where it was located in the early 1980s. As you can tell from the photographs taken at that time (Figures 34 and 139A), a good-sized ebb-tidal delta was located off the inlet's outlet in the early 1980s. Therefore, it stood to reason that the sand that accreted to Seabrook Island in the early 1950s was sand that originally composed the ebb-tidal delta of Captain Sams Inlet in the late 1940s! Once the 1940s inlet was abandoned, shoals of the ebb-tidal delta no longer had the strong ebb currents that formerly exited the inlet throat to maintain them. Therefore, wave-generated currents eventually moved the sand on shore, resulting in the major outgrowth of the shoreline evident on the old aerial photographs.

With that observation in hand, Hayes and his associates (Hayes, Kana, and Barwis, 1980) had a solution for two of the erosion problems plaguing the Seabrook Island development, namely, beach erosion near the golf course and the migration of the inlet into the front third or so of the island. The solution was to "play God" and go back and artificially cut the neck of the spit, which would:

1) Allow the ebb-tidal delta to weld onto the beach in the area of the golf course, solving the erosion there; and

2) Eliminate the threat of the inlet moving into the island for at least another 40-50 years.

This good news was brought immediately to the attention of the engineers concerned with the problem. The response – "Are you out of your mind Hayes? The place where the natural cut occurs is now Beachwalker Park. Can you imagine how

FIGURE 138. Beach Club on Seabrook Island. (A) Eroding dune scarp west of the Beach Club. This scarp retreated over 20 feet during a single high tide. Photograph taken at low tide on 12 October 1974. (B) Revetment composed of riprap constructed to prevent erosion of the type illustrated in A. This photograph was taken in July 1978 a little further away from the Beach Club than the one in A.

FIGURE 139. Beach erosion at the golf course on Seabrook Island. (A) Infrared photograph taken at low tide on 17 October 1978. Note the erosional arc that was developing in front of the golf course (arrow). (B) Photograph taken on 28 January 1983. Arrow A points to sandbag revetment in front of the golf course and arrow B points to riprap revetment in front of houses in the lower right.

many kids would drown in the strong currents in the new inlet throat?"

Okay, so we couldn't cut the spit at the neck. How about going up the spit a shorter distance, say several thousand feet or so? Another problem did exist, however, because the owners of both islands had a claim on the ownership of the spit. Because of the timing of some of the early land grants, the spit was a part of Kiawah on some of the old charts and a part of Seabrook on some of the others (see Figure 128). Anyway, our group (RPI by then) drew up a plan to cut the spit about a third of the way up, and Hayes presented it to anyone who would listen numerous times during trips from Columbia to the

coast. Part of the plan was that the cut would have to be repeated about every 14 years, because of the limited distance up the beach we would be able to relocate the inlet. And, lo and behold, the owners of the two islands agreed with the plan! Oops! Minor detail. Little old lady in tennis shoes sued the management of Seabrook for even contemplating such an insult to the environment, which slowed up the process for another year. Meanwhile, the island lost many more of its valuable acres. During a hearing on Seabrook the plaintiff's lawyer shoved a map into Hayes' chest. Luckily, Hayes was not his usual volatile self that day, and the project was finally approved. However, he immediately made plans for such a scene to not be repeated again anytime soon. How? By getting out of that line of business!

Once the plan was approved, we made the cut for the new inlet and closed the old inlet on 4 March 1983 (see vertical image of the new inlet in Figure 140), using land-based equipment, at the cost of about $300,000. The on-scene management of the project was carried out by Tim Kana, who has published several papers on the project since (e.g., Kana and Mason, 1988; Kana and McKee, 2003).

To make a long story short, almost a thousand feet of beach was created in front of the old eroded scarp in front of the golf course, and, needless to say, the migrating inlet was no longer an immediate threat to the island. The progressive welding of the old ebb-tidal delta sand onto the beach in front of the golf course is illustrated by the photograph in Figure 141A. The inlet was relocated again in 1996 and sand continues to be added to the beaches on Seabrook Island. The width of the added sand in 2006 is illustrated in Figure 141B. Kana (1989)

estimated that over one million cubic yards of sand from the abandoned ebb-tidal delta had accreted to the shore of Seabrook Island after the inlet was relocated in 1983. Those kind of numbers make this inlet relocation project one of the most effective and inexpensive techniques of beach nourishment known to man. If you don't think so, check out the costs for some of the other nourishment projects discussed in this book (this one cost a total of considerably less than one million dollars). As far as we know, this was the first time such a nourishment technique had ever been tried.

Places to Visit

Seabrook Island is a gated community, so access is difficult unless you are a guest on the island. The nearby Bohicket Marina offers boat tours of the area, which we can recommend, based on our experience with the group.

RELOCATION OF
CAPTAIN SAMS INLET
28 MARCH 1983
2000 FT.

FIGURE 140. Vertical image of the relocated Captain Sams Inlet on 28 March 1983. Relocation of the new inlet was completed 24 days earlier, on 4 March 1983.

FIGURE 141. Sand added to Seabrook Island as result of relocating Captain Sams Inlet in 1983. (A) Photograph of Seabrook Island by Tim Kana taken at low tide on 25 February 1986, two years and eleven months after Captain Sams Inlet was relocated. Note the welding and landward migration of swash bars composed of sand released from the abandoned ebb-tidal delta after the inlet was relocated. (B) Photograph by Tim Kana taken on 10 February 2006. The beach in front of the golf course had built out 985 feet in the 23 years since the inlet was originally relocated. It was relocated again in the spring of 1996. Note the position of the 1983 shoreline (compare with photograph in Figure 139B).

EDISTO BEACH AREA

Introduction

The area under discussion in this subcompartment is the shoreline between North Edisto Inlet to the northeast and the mouth of the Edisto River to the southwest, a distance of about 11 miles. Look at the recent satellite image in Figure 142 and you will see that the northeastern two thirds of this subcompartment is composed of a series of undeveloped landward-migrating barrier islands with three minor tidal inlets, at least a couple of which were formed when the landward moving barrier islands intercepted a large tidal creek in the backbarrier area (see also the oblique aerial photograph of these islands in Figure 27). The southwestern one third of the shoreline is a highly developed recurved spit. **Edisto Beach State Park** separates these two segments of the shoreline.

Edisto Beach Recurved Spit

The vertical aerial photograph of the end of the spit taken in 1939 shows that the recurve was completely undeveloped at that time (Figure 143).

FIGURE 142. Vertical infrared image of Edisto Beach subcompartment. Image acquired in 2006. Courtesy of SCDNR.

Carbon-14 dates of mollusk shells collected from the older beaches on the spit show that it has grown to the southwest about a mile in the past 1,500-2,000 years (Stapor and Mathews, 1976). Since their 1976 report, the end of the island has continued to build to the south. Rates recorded by NOAA show that the south end has accreted at a rate of 14.3 feet per year between 1854 and 1956. As development of mostly single family dwellings got underway, some beach erosion was noted on the outer beach (not on the southwest end of the spit), thus an extensive groin field (34 groins) was installed between 1948 and 1975 to stabilize the shoreline position (from NOAA and other sources; see Figure 57). Some of the details of the sediment transport and other present conditions in the area are shown on the map in Figure 144.

Because of the erosion of the beach, several beach nourishment projects have been carried out, including:

- The beach was nourished in **1954 with 850,000 cubic yards** of sediment dredged from the "Yacht Basin," formerly a section of salt marsh landward of the island. The fill consisted of considerable amounts of shell.
- A 12,000-foot section of the beach was renourished in **1995 with 138,500 cubic yards** of sand dredged from an offshore ebb-tidal delta 2,500 feet off of "The Point" at the south end of the island. A number of groins were rehabilitated at that time.

More recently, the town collaborated with

218

FIGURE 143. Vertical image of Edisto Beach taken in 1939. Note the curving beach ridges, which mark the progressive southwesterly growth of the spit.

Colleton County and the state on an $8 million project engineered by Tim Kana's company (Coastal Science and Engineering) to renourish the beach. A dredge spread **850,000 cubic yards** of sand over about 3.5 miles of beach in less than three months in the **spring of 2006**. Oblique aerial photographs taken before and after the nourishment project are shown in Figure 145. The borrow source for this project was a large shoal off the southern tip of the island.

Landward-Migrating Barrier Islands to the Northeast

This shoreline has some of the most dramatic erosion rates found anywhere in the state. The erosion was brought about when the main ebb

channel of the giant ebb-tidal delta at North Edisto Inlet (see Figure 38A) abruptly switched to the north in the mid-1880s, after which time it become a natural groin that trapped in excess of 200,000 cubic yards of sand per year over the next 110 years (Hayes and Sexton, 1983; see Figure 146A). As a result, the barrier island located downdrift of the inlet (Botany Bay Island) has eroded nearly a mile since that time (Figure 146B). The morphology shown by the historic maps indicate that the island was a prograding barrier island in the early 1800s. However, as shown by the oblique aerial photograph in Figure 27 and the image in Figure 142, the island eventually converted from prograding to landward-migrating in nature as it was eroded. A historic summer resort village, Eddingsville, located on the old, prograding barrier island, was completely

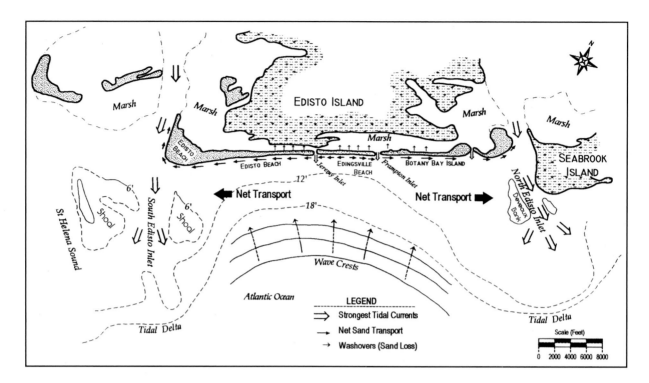

FIGURE 144. Map of the Edisto Beach Area subcompartment showing the dominant longshore sand transport directions. Note that transport is to the north in the northern half of the area, presumably because the large ebb-tidal delta of North Edisto Inlet shields the area from the dominant waves approaching the area from the northeast. Refraction of those waves around the ebb-tidal delta is possibly another factor. At the present time, the landward-migrating (transgressive) barrier island complex know as Edingsville Beach (northern half of area; see also Figures 27 and 142) is moving landward at a rate of about 5 yards per year for reasons given in the text. From Kana, White, and McKee (2004; Figure 6).

washed away. According to Tim Kana, at least one house from the outer beach washed inland to the forest line and became a summer house for a prominent South Carolina family for many years before it burned down in 1972.

Places to Visit

There are rental houses on Edisto Beach, which typically have a flavor of the local culture, because this has been a favorite vacation spot for local South Carolinians for many years. Also, there is the possibility of either launching a watercraft or going on a tour boat from the Edisto Marina, located on the landward side of the recurved spit. The offshore fishing charters out of this marina also have a good reputation, though we have never tried them. The

tidal flats at the entrance to the sound have an abundance of shorebirds and are famous for the large numbers of horseshoe crabs that visit the flats in the spring (discussed earlier, see photographs in Figures 72 and 73). This is also a jumping off place for the incomparable ACE Basin. This part of St. Helena Sound is another one of those places we have visited over 100 times in the past 30+ years, and, as far as a boat ride is concerned, this is the best place in the state to do that, with thousands and thousands of acres of unspoiled marshes, tidal flats, and tidal channels you can float by (not to mention the big flocks of wood storks, snowy egrets, etc. you will see).

The **Edisto Beach State Park** is a popular park with its 1,255-acres of dense maritime forest, expansive salt marsh, and beach. This is about the

FIGURE 145. Aerial views of Edisto Beach taken on 10 February 2006 (A) before a major beach nourishment project and after it in June 2006 (B). Note how the nourished sand had buried many of the groins. Photographs by Tim Kana.

only place in the state where you can drive to and walk on a **landward-migrating (transgressive) barrier island**. Walk north of the camping area for a few hundred yards and you can go right across a shell-rich washover terrace (compare this area with the diagram in Figure 112). Also, if you go at low tide (again!!), you can see the exposed eroding marsh at the toe of the beach face. Great place to see a landward-migrating barrier island. Why is it migrating landward? Rising sea level? Not in this case; it is moving landward because the sand supply was cut off naturally in the mid-1800s (as discussed earlier).

Another aspect of this beach is the steepness of the beachface, because of the abundance of shells in

the sediment. At least a part of that shell material came from the offshore material dredged during the many beach nourishment projects. The relatively steep offshore area makes for larger waves here than you normally see on the South Carolina coast. As a result, the big waves sometimes reflect from the beachface, creating **beach cusps**, and this is the only place in South Carolina accessible by automobile where we have observed rip currents. Beach cusps are evenly spaced triangular projections of coarser-grained sediment (called horns) that project out from near the high-tide line and are separated by eroded out bays, shaped like a half circle, open to the water (see sketch in Figure 147). In this case, the horns will be composed of shell material. Cusps

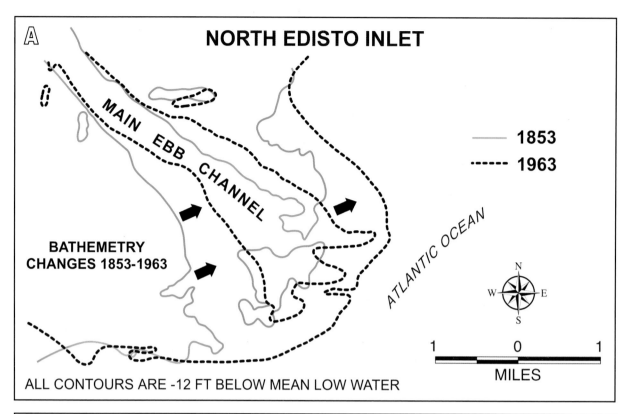

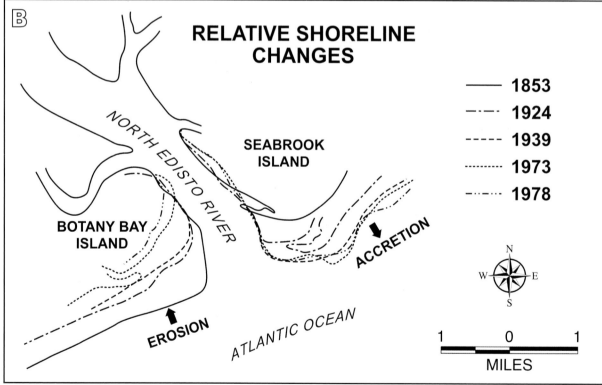

FIGURE 146. Changes at North Edisto Inlet. (A) Illustration of a major shift to the north of the main ebb channel of the ebb-tidal delta. (B) Erosion of the barrier island to the south of the inlet was caused by the shift in the main ebb channel shown in diagram A, which created a new terminal lobe on the delta that trapped a significant percentage of the sand moving south in the longshore transport system.

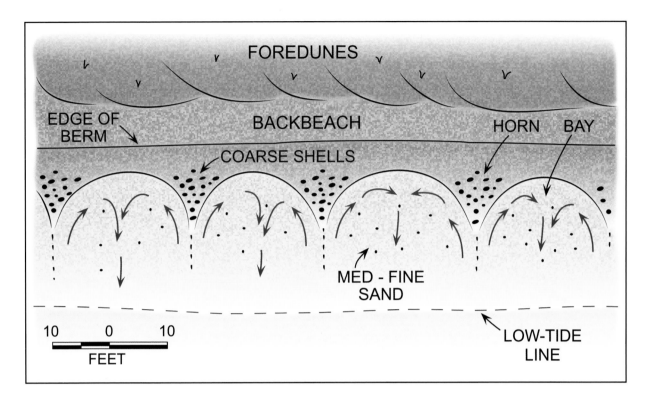

FIGURE 147. General pattern of beach cusps. Arrows show the patterns of flowing water (during high stages of the tide) in the bays between the horns of the cusps.

are thought to form by the same mechanism as rip currents (discussed in Section I). The place to see them is near the groin at the southern end of the beach in the park, where they typically have a spacing of about 20 feet. Beach cusps are very common on beaches with coarse sediment, such as the Great Lakes, California, Maine, and Alaska, but this is one of the few places they commonly occur in South Carolina. Also, do not be too surprised if they are not there on the day of your visit, because they are ephemeral features that come and go in response to changing wave conditions.

Some other amenities of the park include:

- A new interpretive center opened in 2004, with interactive displays and other items of interest.
- Some rental cabins are available.
- A very interesting beach area that offers camping on the beach, with a chance to walk

along a clearly eroding beach. Don't worry, you can camp there. It isn't eroding that fast!
- Great place to collect shells.
- Also a good place to see shorebirds during the migration season.

14 THE LOW COUNTRY

INTRODUCTION

The Low Country of South Carolina is distinctively different from the other three major compartments of the state, with its geomorphic framework of gradual subsidence, considerably larger tides, and its massive estuarine systems. For discussion, we have subdivided this compartment into five subcompartments (Figure 148):

1) **The ACE Basin** – A huge estuarine complex that is fed by three moderately large coastal plain rivers, the Ashepoo, Combahee, and Edisto (hence the name ACE Basin). It is a land of plantations historically known for their rice culture, with many thousands of acres that remain in their pristine natural state. As such, this area is one of the largest unmodified natural areas left on the coastline of the eastern United States.

2) **The Hunting/Fripp Islands headland** – A headland that projects seaward along the southwest side of the lowstand valley complex formed by the three rivers of the ACE Basin. This valley system, which is now filled with salt water, is known as St. Helena Sound. The protruding headland contains several historically prograding barrier islands, some of which are now experiencing erosional episodes.

3) **Port Royal Sound** – Another massive lowstand valley now filled with salt water, which is host to very extensive developments of salt marsh. Access to this area is difficult without the use of watercraft.

4) **The Hilton Head headland** – Another headland to the southeast of a major estuarine complex, Port Royal Sound in this case. The first barrier island in South Carolina to be developed as an upscale resort, Hilton Head Island, is famous for its golf courses, resplendent beach houses, and sometimes crowded beaches, but there are some sites on the island of interest to naturalists. Daufuskie Island, which is accessible only by boat or ferry, is also located on this headland.

5) **The Savannah River Delta** – This delta is the second largest river delta on the east coast of North America. The lower reaches of the delta in South Carolina are mostly inaccessible by car, but the upper delta plain contains the Savannah National Wildlife Refuge, another site no visiting naturalist would want to miss, particularly those with an interest in birds.

The Low Country has a very different ambiance from the rest of the coast, probably because there is so much undeveloped space with extensive swamps and marshes, and many plantations, plus the **tides.**

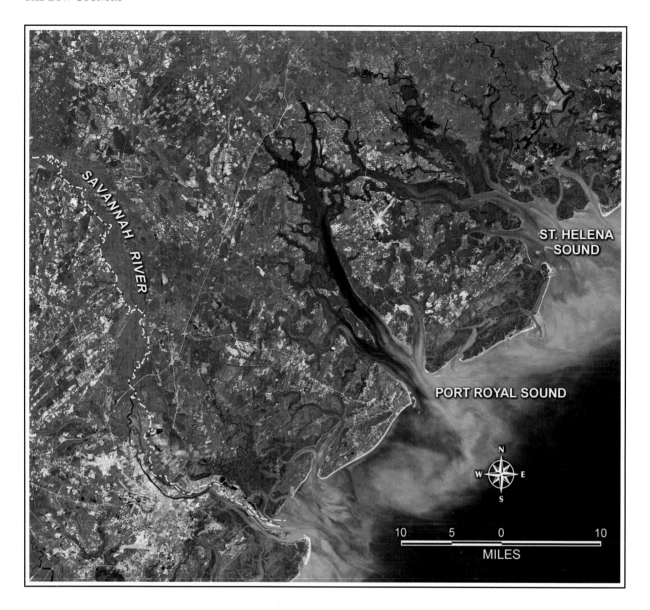

FIGURE 148. Infrared image of Low Country compartment acquired on 23 October 1999, courtesy of Earth Science Data Interface at the Global Land Cover Facility. Note the expansive estuarine marshes and tidal flats that extend up the coastal plain rivers more than 25 miles inland from the coastline.

THE ACE BASIN

Morphology and Origin

The ACE Basin occupies the northern and eastern two thirds of St. Helena Sound, the largest estuarine complex on the southeastern coastline of the United States (see satellite image in Figure 149). This estuary, which is 13 miles wide and 22 miles long, receives freshwater drainage from three main river systems, the Edisto, Combahee, and Ashepoo. As can be seen in Figure 150, the drainages of these three rivers are confined largely to within the Coastal Plain Province of South Carolina. They are all black-water systems, receiving a major component of their flow from the extensive tannic-acid-producing freshwater swamps of the region. The Edisto River, the largest coastal plain river of the Georgia Bight, has an average discharge of around 4,000 cubic feet per second in February-

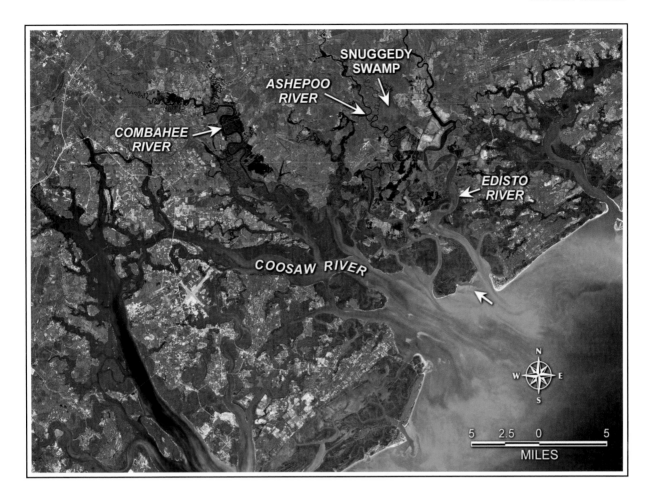

FIGURE 149. Infrared image of St. Helena Sound. Arrow points to the large sand flat in the entrance to the sound pictured in Figure 72. Image acquired on 23 October 1999, courtesy of Earth Science Data Interface at the Global Land Cover Facility.

March. Peak runoff occurs from January through April. The large northwest arm of the estuary, the Coosaw River, receives very little freshwater input.

The largest tides in the Georgia Bight occur in St. Helena Sound and the neighboring Port Royal Sound to the south, exceeding 10 feet on some spring tides. These large tides produce a dominant effect on the estuary, giving it attributes similar to macrotidal estuaries (Hayes, 1976; discussed in Section I, see Figure 11C). The large tides have produced a gigantic shoal outside the entrance to the sound (Figure 149; see also Figure 8), which contains 760,000 acre-feet of clean sand (Hayes et al., 1984; one-acre-foot in this case is a three-dimensional body of sand with an area of one-acre that is one foot thick). Because of the dominance

of tides and the absence of a major source of river sediments (i.e., no river delta), the distribution of sediments and depositional environments shows the following systematic changes landward:

- **Estuary entrance** – Generally sand-rich, high-energy environments (e.g., tidal sand ridges and exposed tidal sand flats; see photograph in Figure 72).
- **Lower estuary** – Transition to quiet water and the zone of the turbidity maximum (e.g., sheltered tidal flats, salt marshes, and complex tidal point bars).
- **Middle estuary** – Transition to fresh water, zone of mud dominance [e.g., extensive brackish water (*Juncus*) marshes, muddy

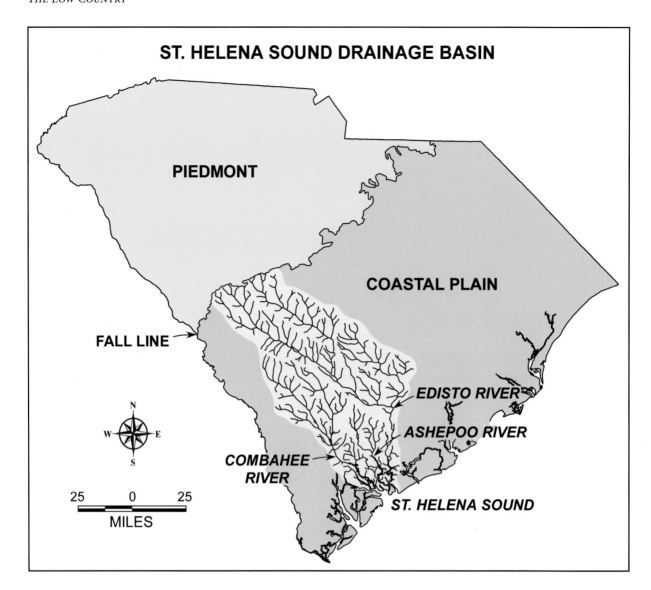

FIGURE 150. Watershed (drainage basin) of the St. Helena Sound estuarine complex. The drainage of the three major rivers of the system, the Edisto, Combahee, and Ashepoo, is confined to the Coastal Plain Province. They are black-water, swamp-draining streams that carry multicycled sand grains reworked from Cretaceous and younger sediments (from McCants, 1982, Fig. 21).

channels, mixed sediments on tidal point bars, and abandoned oxbow mud deposits].

- **Upper estuary** – Freshwater environments dominate (e.g., extensive freshwater peat swamps, sandy point bars in primary streams).

Studies carried out by our group indicate that the depositional history of this estuarine system is one of general infilling with seaward progradation of the surficial habitats in the past 4-6,000 years.

For example, the widespread peat swamp deposits of the upper estuary rest on top of salt-water marsh and tidal flat deposits of the type presently found in the middle and lower estuary.

At the very head of the estuary, the **Snuggedy Swamp** (Figure 149) represents a unique environment with respect to the South Carolina coast. This swamp covers a wide area, but its uniqueness relates to the thick and extensive **peat** deposits it contains, which reach up to 12 feet in thickness in places. The peat is a subtropical reed type that has accumulated *in situ* during Holocene time. In order for sediments to be classified as peat, they must contain at least 75 percent organic material. The peat is typically underlain by a kaolinite-rich clay, which is rooted by salt-marsh grasses at the top. This clay deposit is similar to the underclay deposits associated with coal seams of the Appalachian region. Coal geologists have studied this and other modern-day peat deposits in order to clarify their understanding of how the ancient coal deposits formed.

This particular peat deposit is composed of the remains of plants that consist of many species of freshwater trees and grasses. The peat near the surface is brown to reddish-brown in color and consists of well-preserved plant fragments (termed **hemic**). At increasing depths, it becomes black brown to black and gel-like (termed **sapric**). The basal peats usually have a strong hydrogen sulfide odor. The peat deposits in Snuggedy Swamp are mined and sold commercially.

Figure 151 depicts Staub and Cohen's (1979) hypothesis for the origin of the peat deposits in Snuggedy Swamp. It involves the following steps:

A. When salt water first covered this area about 4,500 years ago, estuarine mud was deposited over the last (Wisconsin) lowstand surface of erosion that formed on Pleistocene sand deposits while sea-level was much lower than it is now. Next, a salt marsh developed over the mud deposit, and then freshwater plants colonized the high spots on the salt-marsh surface.

B. Peat accumulated in areas of freshwater vegetation, causing the development of **peat islands**.

C. The peat deposit continued to grow because the adjacent rivers carried very little sediment into the area.

D. The peat islands coalesced to form a **continuous peat deposit**. Thicker zones of peat occur under

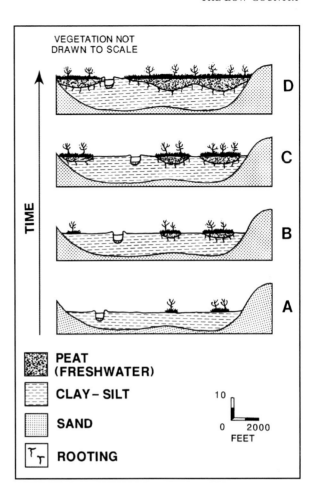

FIGURE 151. Evolution of Snuggedy Swamp peat deposit, according to Staub and Cohen (1979, Fig. 7), from a salt marsh with some fresh water plants on high spots (A) to a continuous peat deposit (D).

the peat islands that were formed earliest.

St. Helena Sound also contains extensive marsh and tidal-flat habitats, but much of the discussion of these habitats in the introductory material in Section I of this book is based on our group and others work in the sound, so that material will not be repeated here.

Places to Visit

As you are driving south from Charleston, there are two possible entries into the northernmost subcompartment of the Low Country, the **ACE Basin**: 1) Where I-95 crosses the Edisto River at Canadys; and 2) Where US 17 crosses the Edisto River. If you are going south along I-95, you may want to check out the **Great Swamp Sanctuary,** which is located on US 17A at the intersection with state road 63 just inside Walterboro. This 842-acre sanctuary, which was once crossed by the historic Charleston to Savannah wagon road, protects an upper bottomland hardwood/swamp forest. The sanctuary contains 1.7 miles of foot trails and canoe access along a 2.1-mile canoe trail. Bicycles and fishing are permitted.

If, on the other hand, you are approaching the ACE Basin along US 17 while driving south, the best way to start your visit is to turn left on Parkers Ferry Road, just before you start to cross the relatively wide flood plain of the Edisto River. The driving route discussed here is shown on the image in Figure 152. As you drive south along Parkers Ferry Road, you will encounter some patches of upper bottomland hardwoods. At 4.4 miles, pull over to view the spectacular avenue of oaks at Prospect Hill plantation (private – no visitors). This avenue is one of the best in the state. The original house was burned by Sherman's troops during his famous march across Georgia and South Carolina in early 1865. This house was built a few years later. How do we know all this? The authors of this book owned this plantation for a few years back in the late 1980s.

At the very end of Parkers Ferry Road you reach **Willtown Bluff**, the site of an old town called Willtown, or New London, established about 1685. Several older buildings are still present at this high bluff on the banks of the Edisto River, a beautiful

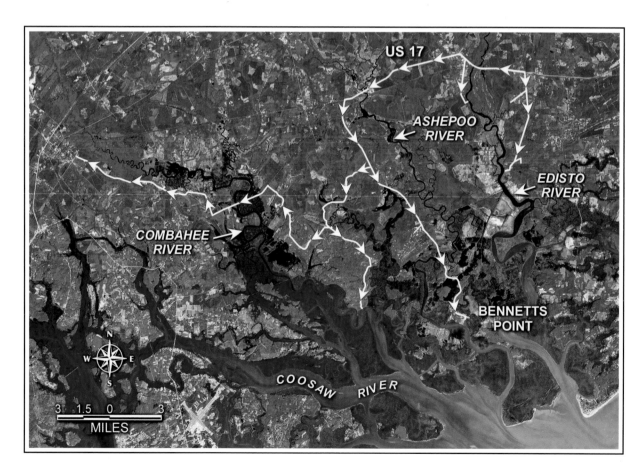

FIGURE 152. Infrared image of part of the ACE Basin showing the location of the recommended driving tour of the area. Image acquired on 23 October 1999, courtesy of Earth Science Data Interface at the Global Land Cover Facility.

black-water river that is the boundary between Colleton and Charleston counties. The stately grounds on the bluff contain imposing old live oak trees with abundant **Spanish moss**. A boat ramp at the end of the road maintained by Charleston County Parks and Recreation Commission provides boating access to the upper reaches of the Edisto River estuary, one of the most picturesque in the state.

After you leave the boat ramp at Willtown Bluff, take a right on county road 55. This is a dirt road for a half mile and from there on it is paved. Go about a mile on the paved road and then take a right on county road 346. As you drive down this road, you are headed towards the Grove Plantation, the Headquarters of **Ernest F. Hollings ACE Basin National Wildlife Refuge**. This road is located on a Pleistocene upland with mixed pines and hardwoods. Upper bottomland hardwoods with dwarfed palmetto occur in some of the topographically low areas as you get closer to the plantation. The house is a couple of miles in. Gates open 7:30 – 4, Monday through Friday, and the grounds are open daylight only. You can park outside of the gates. There are walking trails around a pond and on other parts of the property. The grounds are gorgeous, with big live oaks inhabited by an abundance of song birds. However, there is no avenue of oaks. Grove Plantation has one of the three plantation houses in the whole ACE Basin that Sherman's troops did not burn, so this is a rare opportunity to view the architecture of one of the original plantation houses (see photograph in Figure 107B).

At the headquarters, which is located inside the main plantation house, you can obtain an abundance of printed information as well as view some displays. Two printed items you wouldn't want to miss are: 1) a bird list for the refuge; and 2) a large fold out map that gives recommended driving routes and other information about the ACE Basin.

Now we will continue the description of our recommended **driving tour of the ACE Basin**

(Figure 152). To proceed with the tour, drive back the way you came down Parkers Ferry Road, saying goodbye to Prospect Hill plantation as you pass it. A few hundred yards short of Highway 17, turn left on the road to **Penny Creek Landing**. This high-water boat landing is located on a delicate little black-water creek that is still tidal, although it is located near the upper reaches of the Edisto River estuary. This would be a good place to launch a canoe or kayak for a paddle through these attractive wetlands. It is about a mile and a half down to the main channel of the Edisto River. This is no doubt a prime birding spot in the spring.

Back on US 17, turn south and drive across the wide coastal plain of the Edisto River. Pull over on the right soon after you cross the bridge over the river into the parking area for the **ACE Basin Gateway Edisto Nature Trail**, which is maintained by Mead Westvaco. A description of the Basin on a sign at the trail entrance follows:

> "*The ACE Basin consists of approximately 350,000 acres of diverse habitats, including pine and hardwood uplands, forested wetlands, fresh, brackish and salt water tidal marshes, barrier islands and beaches. The Basin's unique estuarine system, the largest of its type in the state, provides a variety of habitat for a rich diversity of finfish and shellfish resources. The Basin hosts a wealth of wildlife resources, including endangered and threatened species, such as the bald eagle, the osprey, the wood stork, loggerhead sea turtle and the short-nosed sturgeon, and offers a variety of recreational uses. The mission of the ACE Basin Project is to maintain the natural character of the Basin by promoting wise resource management on private land and protecting strategic tracts by conservation agencies. A major goal of the protection effort is to insure that traditional uses such as farming, recreational and commercial fishing and hunting will continue in the area.*"

After leaving this area, continue south on US 17. In Jacksonboro, turn left on the Hope Plantation road. **West Bank Landing** on the Edisto River is located about 1.5 miles down the road. This is a serviceable boat ramp with a large parking area that provides a notable view of the Edisto River in the upper reaches of the estuary. Across the channel you can view an extensive marsh, tan-colored in the fall and a radiant green in the spring and summer, backed by swamp trees. A little further down the road from the boat ramp, it is closed to public traffic.

Back on US 17, continue south until you cross the Ashepoo River. On the left as soon as you cross the bridge is Jill's Fish Camp, where you can rent canoes. The ramp fee is $5 at this very popular site. About a mile further south, turn left on Bennetts Point Road and begin a 13-mile drive to Bear Island Wildlife Management Area (WMA). This is a quintessential low-country drive. Every mile or two you come upon an entrance to something they are calling plantations over near the Ashepoo River on your left. Many of these large land holdings have some kind of easement agreements with the ACE Basin Authority through the Nature Conservancy and the U.S. Fish and Wildlife Service.

We recently made this drive on a Thanksgiving weekend and the fall colors were very striking, even for someone who lived in New England for nearly 10 years. About 4 or 5 miles down the road, a **rice pond** is present on the right side of the road, and there seems to be always someone fishing in the drainage canal below the road. After about 10 miles you pass the entrance to Airy Hall plantation on the left. This superbly landscaped horse farm (also private) is on the bank of the Ashepoo River.

Continuing south for a few hundred more yards, you will approach the bridge across the Ashepoo River. Park on the right side of the road and walk up on the bridge, from which you can view the vast *Juncus roemerianus* **marsh** in the upper limits of this arm of the St. Helena Sound estuary. A little fringe of *Spartina cynosuroides* occurs right along the edge of the main channel. After you cross the bridge

driving south, you pass through several hundred yards of a broad monospecific *Juncus* marsh!! You don't see many of these by driving. A "wildlife observation" area with a picnic table is located just beyond the edge of the marsh (on the right). This is a good place to view the big flocks of **white ibises,** wood storks, common egrets, and other wading birds that populate this area. We almost always see bald eagles in this part of the basin as well. A little further along is the **Bear Island WMA** entrance, where general visitation is from 1 February to 14 October (open Monday-Saturday daylight hours). Hunting by permit only. Fishing and crabbing is limited to between 1 April and 30 September. There are some **excellent walking trails** in this WMA along the edges of the old rice ponds and so on.

The web page for the Carolina Bird Club describes this WMA as follows:

> With 12,000 acres of managed wetlands, scattered stands of pines and agricultural fields, Bear Island attracts waterfowl, bald eagles, wading birds, shorebirds, raptors and songbirds. This area includes an observation platform, roads for driving, and miles of dikes for walking. Truly a must-see place. Look for bald eagle, great horned owl, wood stork, and black-necked stilt. .... A good day at Bear Island can yield over 80 species. Primitive rest room and picnic facilities are available near the entrance.

From the WMA you might want to continue on to the south end of the road at Bennetts Point, which is not exactly a hot spot for naturalists but is an interesting cultural experience. As you drive along through this area you will notice what appear to be randomly distributed islands in the marsh that are host to a climax maritime forest (live oaks, loblolly pines, palmettos, etc.). These are topographic highs composed of Pleistocene sediments left as a result of erosion by the distributaries of the three main rivers in the ACE Basin when sea level was lower.

As sea level rose to its present high, these erosional remnants became surrounded in many places by marsh grasses.

To continue our auto tour of the ACE Basin, drive back up Bennets Point Road for about 6 miles to where you will see a sign pointing the way to **Donnelly WMA** along Blocker Run Road (turn left).

The web page for the Carolina Bird Club describes this WMA as follows:

The 8,000 acres of Donnelly contain one of the East Coast's largest undeveloped areas of estuaries and associated wetlands with pine and pine-hardwood uplands, bottomland hardwoods, managed wetlands and agricultural fields. ... Look for bald eagle, osprey, wood stork, purple gallinule, anhinga, wild turkey, and hope for black-bellied whistling duck. Waterfowl, waders, raptors, and passerines make Donnelly WMA an excellent location for seeing a variety of birds.

This WMA is open for visitation 8-5 pm Monday-Saturday, closed Sunday and during special hunts. As you drive west along Blocker Run Road, you first go through managed pine forests. There are some obvious hiking roads through the forests, though they are not designated as such. In about 2.5 miles, you encounter some buildings with open fields where they are growing sunflowers and corn, presumably food for the wildlife in the area. A little further along is the **Savage backwater and restoration project**. Beautiful black water, a good place to fish for brim!! Gallinules and lots of wood duck houses are present in this alluring spot.

Keep following the main road to the west until you come to a dam over the headwaters of Old Chehaw River. The road goes over the dam. We have visited this pond several times and have never failed to see a multitude of alligators (Figure 153) and numerous birds, including plentiful cormorants, great blue herons, and white pelicans, as well as the occasional bald eagle.

Continue on over the dam and along the road until you come to a T intersection. Turn left and follow this dirt road for a mile or so until you come to a paved road (Wiggins Road), where you turn left and soon come to the turn to the **Chehaw River Access,** a well-maintained boat ramp with a spacious view of the marsh complex at the head of this estuary.

After a respite at the ramp, where you should see more interesting birds, return to the paved road and continue to the south (left). This drive is a good opportunity to see the undeveloped upland part of the ACE Basin. At 1.5 miles down Wiggens Road, the pavement ends and at 2.5 miles is Wiggens, a private inholding in this mostly unpopulated upland of pines and hardwoods.

Continuing along the dirt road for about 5 more miles you come to **Field's Point Landing**, a boat ramp on a moderately high bluff in the Pleistocene upland on the shore of the lower portion of the Combahee River. The end of this road at this superb middle/lower estuary complex is the second most seaward point you can reach by driving in the ACE Basin. Bennetts Point is the most seaward one, so this road tour has taken you to both of those spots.

As you head back up the road, within a couple of miles you will pass through a stand of **longleaf pines** (*Pinus palustris*). This tall, stately pine once covered 30-60 million acres in the southeastern United States coastal plain. Logging and land clearing have greatly reduced its range. You may want to get out of the car and observe the two defining characteristics of this variety of the pine family: 1) Very large cones 6-10 inches long (some people find these to be unique collector's items); and 2) Long needles between 8-18 inches long, which are sheathed in fascicles of three each.

Drive back to the Old Chehaw River Access and keep going along the paved road toward the northwest (do not turn onto the dirt road you came in on). We think this is also a pleasing drive,

FIGURE 153. Old Chehaw pond in the Donnelly Wildlife Management Area. (A) General view of the pond from the road on the dam. Note the alligators in the upper right. (B) Closer view of two of the gators in the pond. Photographs taken on 24 November 2006.

a mosaic of land patterns with minor development. The variety of the pine forests and the older hardwood stands helps maintain your interest along the drive. This stretch of the road provides views of a great variety of trees, which are especially appealing around Thanksgiving time because of the fall colors. A couple of immaculately landscaped plantation entrances are located about half way out to US 17.

Once back to US 17, turn left and drive south until you cross the Combahee River. You may want to check out the boat ramp on the right after you cross the river (Steel Bridge Boat Landing). Even though you are far up the Combahee estuary at this point, the expansive marsh is primarily composed of smooth cordgrass (*Spartina alterniflora*). Salt marsh this far inland is a result of the large tides in this part of the Georgia Bight and the relatively small freshwater input by the Combahee River (winter-time flow of around 500 cubic feet per second), perhaps the most beautiful of the three main rivers in the ACE Basin.

After you leave the boat ramp, or continue south from the bridge without stopping, turn right on River Road and drive north, our favorite auto tour in all of the state. This short drive, which ends at Yemassee, has a little bit of everything – pecan orchards, elegant plantation houses and grounds, and bottomland hardwoods, and even a drive by *Aldebrass*, Frank Lloyd Wright's southern estate (open only rarely to visitors).

About half way to Yemassee, keep an eye out for the sign for **Sugar Hill Landing** and drive to it. This ramp, a choice spot to launch a canoe, kayak, or jon boat, is on a side channel that leads out to the main channel of the Combahee River. This boat ride is a spectacular trip through freshwater wetlands at the very head of the Combahee estuary.

Continue the drive up River Road until you reach Yemassee, the end of this auto tour through the ACE Basin. No matter how you leave or enter the place, you will always see a demarcation in the charm of the area as soon as you leave or enter the ACE Basin, our favorite part of the South Carolina coast.

THE HUNTING/FRIPP ISLANDS HEADLAND

Morphology, Origin, and History

Figure 154 gives a satellite image view of this headland, which is sandwiched between two large, flooded lowstand valley systems, St. Helena Sound to the north and Port Royal Sound to the south. After sea level rose at the end of the Pleistocene Epoch, six barrier islands formed on the outer margin of this projecting headland – two larger ones, Hunting and Fripp Islands, which have road access and coastal residences in places, and five smaller ones, Harbor, Pritchards, Capers, St. Phillips, and Bay Point. Of the smaller islands, only Harbor Island is developed at this time. A significant portion of Hunting Island is a state park, another one of those sites no visiting naturalist would want to miss.

The huge ebb-tidal delta/estuarine entrance shoals, as well as the strong tidal currents, associated with the two adjacent sounds have an important impact on the morphology and coastal processes of the adjacent islands, especially Harbor and Hunting Islands to the north and Bay Point and St. Phillips Islands to the south. The ebb-tidal delta of Trenchards Inlet, which separates St. Phillips and Capers Islands, is molded into the massive estuarine entrance shoal of Port Royal Sound. The huge ebb-tidal delta of Fripp Inlet, pictured in Figure 155, is very conspicuous, dominating that part of the nearshore zone. There is no evidence whatsoever of a flood-tidal delta at this inlet, one we sometimes use as a model of the extreme end member of ebb-dominated tidal inlets (see discussion in Section I). Fripp Island, and its associated ebb-tidal delta, project further out into the Atlantic Ocean than any other barrier island system in the state.

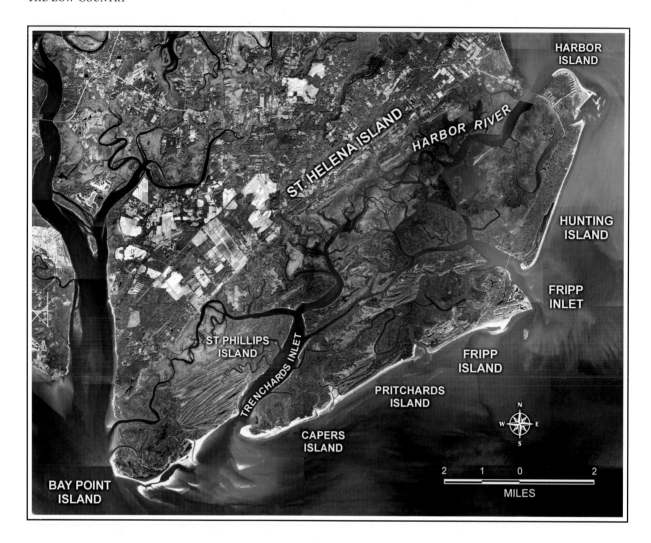

FIGURE 154. Vertical infrared image of Hunting/Fripp headland. Image acquired in 2006. Courtesy of SCDNR. The sharp break between St. Helena Island and the salt marsh marks location of the shoreline during the last high sea level of the Pleistocene Epoch. All of the marshes, tidal flats, and barrier islands seaward of that line were formed during the present highstand of sea level (approximately the last 4,500 years).

Look at the image in Figure 154 and you will see the following principal aspects of this headland's morphology:

- The highly developed, linear upland in the northwest portion of the image, St. Helena Island, an older Pleistocene highstand deposit.
- A complex network of tidal channels, salt marshes, and tidal flats are located immediately seaward of St. Helena Island. Note the abundance of tidal flats between St. Helena Island and Fripp Inlet (Harbor River flats). Centrally located

tidal flats like this are common in the central regions behind barrier islands from here north to New Jersey (see Figure 75A). As noted in Section I, our hypothesis for the cause of these flats is that the suspended sediment coming into the backbarrier region from offshore, in the absence of freshwater streams in the area, is trapped by salt-marsh vegetation before the flood-tide waters reach the point where the "tides meet," also known as the head of the two adjoining watersheds. In this case, the mud flats are located where the water coming into

FIGURE 155. The exceptionally large ebb-tidal delta at Fripp Inlet. The central channel (main ebb channel) is dominated by ebb currents. Infrared photograph taken at low tide in December 1979.

the backbarrier through the main channel of Trenchards Inlet to the southwest meets the water coming in through Harbor River to the northeast.

- An abundance of prograding beach ridges on the barrier islands along the outer coast and relict beach ridges surrounded by salt marsh. Fripp Island conforms closely to the drumstick model (see Figure 29).

Hunting and Fripp Islands will be discussed in considerable detail below. Some notes on the other islands follow:

- Capers and Bay Point Islands, which can be reached only by boat, are both rather small and are obviously erosional; therefore, development

of them would be quite risky.

- Harbor Island, which is much more accessible, is already heavily developed. According to DHEC's March 2006 State of the Beaches Report, *"the shoreline of Harbor Island, while being very dynamic, is generally accretional in the long term."* No doubt, some of the sand deposited on Hunting Island during several beach nourishment projects ends up on Harbor Island.
- Pritchards Island, another completely undeveloped island, was donated to the University of South Carolina Development Foundation in 1983. Currently, the island, which is managed by the University of South Carolina Beaufort, is used for education, conservation, and research purposes by the University, other

state institutions, and the general public (www. uscb.edu/).

- St. Phillips Island, which is approximately four miles long and two miles wide, is private, and according to the National Park Service, "*it is unique among the barrier islands of Georgia, South Carolina and northern Florida, because it exists in a nearly undisturbed state with minimal development and past consumptive use.*" They further state that "*it is also unique to the entire Atlantic Coast for the pronounced multiple vegetated beach dune ridges found here.*" Those beach ridges are pretty spectacular all right (see Figure 154), but this last statement is debateable. What about Bull and Kiawah Islands?

Places to Visit

The two key areas to visit on this protruding headland are the picturesque coastal town of **Beaufort** (pronounced byou' furt) and **Hunting Island State Park**. If you are driving south on US 17, to get to Beaufort turn left on US 21 at Gardens Corner and continue south to the bridge crossing the Coosaw River. The estuary at this point contains numerous circular mounds of salt marsh (*Spartina alterniflora*) with abundant edges; therefore, the marsh grass is tall along those edges. These semicircular patches of marsh grass appear to be growing on old oyster mounds, where the oysters have apparently accreted their own mud (feces and pseudo feces). The mud flats are very extensive in this area, with limited amounts of marsh, probably because of the small sediment input by the Coosaw River. Consequently, this is probably another example of mud flats where there has not been enough sediment to build the tidal flats up to near mean sea level where the marsh grasses can grow on them.

Approaching Beaufort, near Grays Hill, you see more wide expanses of lower salt marsh (*Spartina alterniflora*) with scattered mud flats and oyster mounds. Much of the Beaufort River, which meanders around the waterfront of Beaufort, is partially filled in with salt marsh. You will probably also observe the salt-water dominance of the marsh plants in this middle part of the Low Country, an area not influenced by the larger freshwater rivers located to the north and south.

Parts of the town of Beaufort are built on high bluffs in Pleistocene sediments. These bluffs have steep scarps in places that were carved by either migrating tidal channels or wave erosion at high tide when the sea first reached its present level around four to five thousand years ago.

While in this area, you may want to walk along the waterfront in Beaufort, a miniature Charleston with a variety of shops and restaurants. A worthwhile experience would be to extend your walk to include a jaunt across the bridge over the Beaufort River (US 21 bridge), which contains a walking path. This trek affords some distinctive views of the town as well as of the estuarine complex itself.

The essence of this area is captured in the celebrated book, *The Prince of Tides*, by the well-known South Carolina resident, Pat Conroy. The protagonist in the novel, Tom Wingo, grew up on the South Carolina coast near a town called Colleton, which was obviously patterned after Beaufort (the county seat of Beaufort County). This is how he defined the area we are now trying to picture for you:

> *To describe our growing up in the lowcountry of South Carolina, I would have to take you to the marsh on a spring day, flush the great blue heron from its silent occupation, scatter marsh hens as we sink to our knees in mud, open for you an oyster with a pocketknife and feed it to you from the shell and say, "There, that taste. That's the taste of my childhood." I would say, "There, breathe deeply," and you would breathe and remember that smell for the rest of your life .... My soul grazes like a lamb on the beauty of the indrawn tides.*

We left out some words. Everything he writes is a helix of multiple nuances.

Tom Wingo had an eccentric grandmother, Tolitha, who took multiple sabbaticals from a number of husbands to roam the world for a while, before returning home to his fanatically religious grandfather, who walked along the streets of Colleton bearing a huge cross every Good Friday. Here is how Tolitha described the lowcountry:

But all the places I saw in Europe, Africa, and Asia, some were so beautiful it would make you cry. But none of them could make me forget the marsh and the river right out yonder. The smell of this place rides in the bones wherever you go. I don't know if that's a good or a bad thing.

The trip on out to Hunting Island State Park via US 21 involves crossing a few miles of the width of **St. Helena Island**, which is made up of a prograding series of Pleistocene beach ridges formed about 100,000 years ago (over an order of magnitude more than 6,000 years, or did we say that already?) during the highstand of sea level that preceded the last glaciation. On St. Helena Island, there is a fringe of brackish *Juncus* marsh around the shoreline in places, probably due to freshwater runoff and ground water seepage from the Pleistocene highland. But as you cross the bridge to Hunting Island, you see only a very broad salt marsh (*Spartina alterniflora*). There is not a bit of *Juncus* around this causeway, which is very different from the Pleistocene St. Helena Island with its freshwater lens. If you visit this area at low tide, you will note some wide tidal flats and marshes with abundant birds. Typically, white ibises and other wading birds are the most common species present.

From the bridge, you can see all the way to Edisto Beach, the northern boundary of the Low Country compartment, a great view. In this area you can also observe thousands and thousands of acres of dwarfed *Spartina*, one of the largest expanses of this you can see anywhere by driving on this coast. After you cross the bridge, pass by the development on an upland a few hundred yards southwest of the main body of Harbor Island, and

drive south along the landward side of the Hunting Island, where you can survey a superb panorama of the wide salt marsh, especially in the early morning (looking away from the sun).

At the entry to Hunting Island State Park, pay the fee of $4 per adult and drive through the mature climax maritime forest out to the beach, where there are spacious parking areas, picnic tables, and rest rooms. This readily accessible bathing beach is very flat, being composed of very fine-grained sand, with pines right at the high-tide line (see photograph in Figure 156A). A bit further south from that parking area, there is a lagoon on the landward side of a narrow beach ridge located just back of the high-tide line. When we visited the park on Thanksgiving weekend in 2006, the more southerly portion of the beach was eroding. The intertidal beach in that area was very flat, with a bit of a scarp at the high tide line that had a few trees sticking out of it (Figure 156B).

On the other hand, a broad welded berm with shell hash in the swale was located about half way between the swimming beach and the erosion zone at the south end of the island. This was the result of the fact that a subtle bulge (rhythmic topography; Figure 22C) was projecting out into the surf a little ways to the north of that point. A snow fence has been established to build up the foredune in that area.

This beach has been the site of a number of beach nourishment projects, the most recent one being done in the summer of 2006. This latest project is discussed as follows on the web page of Coastal Science and Engineering, LLC, Tim Kana's firm based in Columbia, South Carolina, which, as noted earlier, has been involved in many projects of this type, including this one:

"Hunting Island has been a poster beach for opponents of nourishment because of the frequency and number of projects attempted. It is one of the places where one can truly say this project will only last about three years. But after all, should more longevity be expected for

FIGURE 156. The beach at Hunting Island. (A) View looking north at low tide from near the main parking area. (B) View looking south of the eroding beach near the south end of the island. Both photographs taken on 26 November 2006.

a beach that has lost up to 25 feet per year since the 1940s?

"Hunting Island's erosion rate is ten times faster than Myrtle Beach's and much of South Carolina's developed coast. For years, it was the only beach in South Carolina receiving regular infusions of sand. Eight projects in 38 years added nearly 5.5 million cubic yards. Yet even with all this sand concentrated along the 4-mile-long island, today's shoreline is hundreds of feet landward of the 1960 foredune.

"The 2006 project differs from previous nourishments because it will be followed by construction of up to six groins. Strategic areas where groins will be placed are the campground, North Beach at the lighthouse, and the South Beach recreation area, which received all the 2006 nourishment sand. In some places, the new dry beach is 175 feet wide. Like projects before it, these nourished areas will erode rapidly until groins are constructed (expected in fall 2006). With groins in place, the erosion rates will lessen significantly, but will not stop completely. Hunting Island will require future nourishment even with groins in place – just not as much or as often."

We did not see any groins when we visited the beach on Thanksgiving weekend in 2006, but Tim Kana tells us that they were under construction in early 2007. A photograph of the part of the beach just described taken by us years before the project (January 1975) and one taken in June 2006 (after nourishment) are shown in Figure 157 (see also the book cover). It is clear from comparison of the two photographs that the front of the island had eroded significantly during that 31+ year period.

The beach is not the only interesting part of the island to visit. We recommend also a driving tour of some of the back roads through the dense forest. You will note that the roads pass over a number of ridges and swales, which developed as a result of the prograding growth of the island in the early stages of its development. The ridges are remnants of the lines of old foredunes formerly located just back of the beach.

The visitor's center, with its exhibits and literature, is another place you shouldn't miss. Like most of the natural areas of the coastal region of the state with freshwater habitats, alligators are present in this area. A sign by the visitor's center provides this interesting and succinct discussion about the gators:

"The American alligator belongs to the order Crocodila, which contains 21 living species worldwide. Growing to 18 feet, this large reptile's body form has changed little in 180 million years. The American alligator is the only crocodilian native to our area. Because of the commercial value placed on their skin, their numbers were seriously reduced in the early to mid 20th century. Today, because of strict protection measures, they have made a rapid recovery. The female constructs a mound of vegetation in which she deposits about 30 eggs. Heat caused by the decomposition of the organic material incubates the eggs. The female guards the eggs from potential predators such as raccoons. The young, which are most vulnerable to predators after hatching, hide among the edge of vegetation in shallow water. Remember that alligators are large, powerful and able animals. Respect them as you would any wild creatures. Adult male alligators frequent open bodies of water to feed while the females live in shallow marsh habitats. These predators eat a variety of animals, including fish, snakes, birds, small mammals, carrion, and other small alligators. While in the water, powerful strokes of the alligator's tail allow it to quickly approach its prey. Although slower on land, they can move with startling speed for short distances. Courtship and breeding occur in late April and May. About 65 days after the eggs are deposited, the 9-inch long hatchlings emerge."

FIGURE 157. Hunting Island. (A) Oblique view looking north at low tide on 30 January 1975. (B) View a little further north along the beach taken soon after a beach nourishment project (on 9 June 2006). Photograph by Tim Kana. The arrows, which point to the same location on both images, illustrate the extreme amount of erosion that had occurred on this island within the 31 years between the times the pictures were taken.

Another sign at the visitors center describes duckweed, one of the world's smallest flowering plants, which provides food for some wetland animals and covers the pond surfaces reducing evaporation rates during droughts and preventing the ponds from going dry. Appropriately, the pond next to the visitor's center is usually covered with duckweed.

Of course, further information about this state park can be obtained from the Internet. The following information on Hunting Island as a nesting ground for the loggerhead sea turtle, which is threatened throughout its range, was provided by www.seasidegetaways.com:

- *These turtles come onshore on the island in June, crawling up the beach to above the high-tide line where they dig a hole and lay 100-160 eggs and bury them before returning to the sea.*
- *At night, Hunting Island personnel, along with volunteers, comb the beach for new nests. They then uncover the eggs and take them to a hatching facility located on Hunting Island where the eggs can hatch and the young turtles are safe from the many predators.*
- *After hatching, the young are released to begin their journey to the sea.*
- *If you should find a nest, it is important not to disturb it. Report it to a park ranger.*

That same source also notes that *"sand dollars and sea shell collecting on Hunting Island beach is very popular. The best time to find them is during low tides. (No kidding, go to the beach at low tide!!). The lower the tide the better* (hmmm.). *Often after a storm or spring tide, many sand dollars may wash up on the beach."*

While you are in the park, by all means pay a visit to the **Hunting Island Light Station** and climb up to the top (if it is open) for a fabulous panoramic view of the whole island. This light guided mariners along this stretch of coast for many years. The structure you will climb to the top of (hopefully), however, is not the original lighthouse, which was constructed of brick and completed in 1859. That first structure was demolished by Confederate troops during the early days of the Second Civil War. Does that sound familiar? Remember the lighthouse on Morris Island? The present light-house, built of interlocking cast iron plates, was completed in 1875. To give an idea of how much erosion has "plagued" Hunting Island over the years, the reconstructed light station was originally located on the northern end of the island. Soon after it was set up in 1875, severe erosion soon threatened the station. By 1888, high tide was within 35 feet of the keeper's house and the lighthouse was relocated the following year to where it is today (well back of the high-tide line). In its renewed position, Hunting Island Light continued to aid navigators until 1933, when a buoy took over its function.

Why this island is eroding so much faster than most of the other barrier islands in South Carolina is an interesting problem, but one we have never had the opportunity to study in any detail. One key factor, no doubt, is the presence of the large, complex shoal at the entrance to St. Helena Sound, which is located just to the north of the island. Currents fueled by the largest tides in the Georgia Bight flow in and out of numerous major channels through these shoals. The currents themselves do not erode the island, but sand removed from the beach by wave action during storms could easily be

swept away by the strong currents. The constantly shifting main axes of the channels is also no doubt an important factor. Sand shoals associated with the channels could also play a role in focusing refracted waves against the north end of the island.

When you exit the main entrance to the park, it would be well worth your while to turn south (left) so you can enjoy another inspiring view of the marsh at a little pullover a ways on past the entrance to the park. This marsh is mostly *Spartina alterniflora* with just a little rim of *Juncus* up against the trees.

By all means be sure you don't miss the **marsh boardwalk**, located a little ways further towards the south end of the island (on the main road). This boardwalk is one of the best places in the whole state to view up close the different plant species of the coastal marshes. A great place to see the **high salt marsh** occurs where the boardwalk passes over the edge of a hammock. Note the distinct zones of *Borrichea* and *Salicornia*, with a little bit of *Spartina patens* mixed in. Refer to Figures 69, 70, and 71 for the marsh zones and species. In the middle of the walk, the boardwalk passes over a tidal flat with abundant **fiddler crab** burrows. The boardwalk ends at a medium-sized tidal channel that snakes through the low marsh (*Spartina alterniflora*). As you return to your car, don't forget to listen for the abundant clapper rails making their *clickety-clack* calls throughout the marsh (see Figure 158 for photographs of this marsh and tidal creek).

A sign along the walk points out the importance of the salt marsh ecosystem:

"The important role of the salt marsh is to serve as a nursery for ocean-dwelling species. Many newborns are hatched in offshore waters. Lucky newborns will be carried to the marsh on the tides. There they thrive for weeks or even months until they mature and begin their migration out to sea. A healthy salt marsh is vital to these newborns' survival. Unfortunately, many of our salt marshes are threatened by non-point-source

FIGURE 158. The salt marsh on the landward side of the southern end of Hunting Island (a few hundred yards north of the bridge across Fripp Inlet). (A) *Spartina alterniflora* salt marsh. View looking north from boardwalk. (B) Tidal channel through salt marsh at the end of the boardwalk. Photographs taken on 26 November 2006.

pollution, such as runoff from septic tanks, agricultural land, storm drainages, streets, and yards. Water quality can be improved and healthy marshes protected if we eliminate non-point-source pollution."

The last stop on the tour of Hunting Island should be at the fishing pier at Fripp Inlet. Walk out to the end of the pier and observe the tremendous amount of sand coming around the south end of Hunting Island, which is no doubt derived from the many beach nourishment projects. However, according to Tim Kana, the bulk of the nourished sand is transported to the north where "Johnson Creek inlet and Harbor Island have gotten most of the eroded sand." The fishing pier is also a good place to see some ocean birds, maybe. And go fishing if you so desire. The small nature center on the pier is also worth a visit.

While at the pier, you can look seaward along the main ebb channel of the Fripp Inlet ebb-tidal delta, which is pictured in Figure 155.

Fripp Island is a private, gated community. Housing is available for vacationers, and if you visit, you would have access to some interesting beaches, golf courses, and fishing opportunities. Some general facts about the island include:

- The high-tide line of much of the length of the island is armored revetments.
- A large swash bar derived from the huge sand mass of the ebb-tidal delta came ashore several hundred yards south of Fripp Inlet in 2004 *"greatly widening the beach"* (DHEC). Although the local observers expressed some surprise over the appearance of this "exotic" bar, this type of bar welding to the beach is common on the updrift end of drumstick-shaped barrier islands such as Fripp. Refer to Figures 56A, 56B, and 129 for examples of this type of bar welding on the northeast end of Kiawah Island and Figure 119B for one on the northeast end of the Isle of Palms.

- Otherwise, the beaches have been generally stable for the past several years, presumably because of the heavy armoring of the shoreline.
- A bridge built in 1961 connected Fripp with Hunting Island, and within 10 years, the development of the island was in full bloom.
- Fripp has a year-round population of 887 residents (2000 census), but in summer months, the island's population can rise to about 2,000 to 3,000 visitors.
- It is seen by some as an alternative, affordable, and less "touristy" destination compared with its large neighbor to the south, Hilton Head Island.
- Several movies have been at least partially filmed there, including *Forest Gump* and *The Prince of Tides*.
- The aforementioned author of the book, *The Prince of Tides*, Pat Conroy, is a resident of Fripp Island and he has written the introduction to a short book entitled *Fripp Island: A History* (by Page Putnam Miller).

PORT ROYAL SOUND

Morphology and History

A few of the major morphologic aspects of Port Royal Sound clearly visible on the satellite image in Figure 159 include the following:

- A huge estuarine entrance shoal composed of sand, built by the strong ebb-tidal currents in the sound, projects far out onto the continental shelf. See also the image of the continental shelf in Figure 8.
- Port Royal Sound is located in a very large lowstand valley with little freshwater inflow at the present time, and it is saline over most of is extent.
- The main channel of the sound is straight and deep. Why it is so straight relative to other major

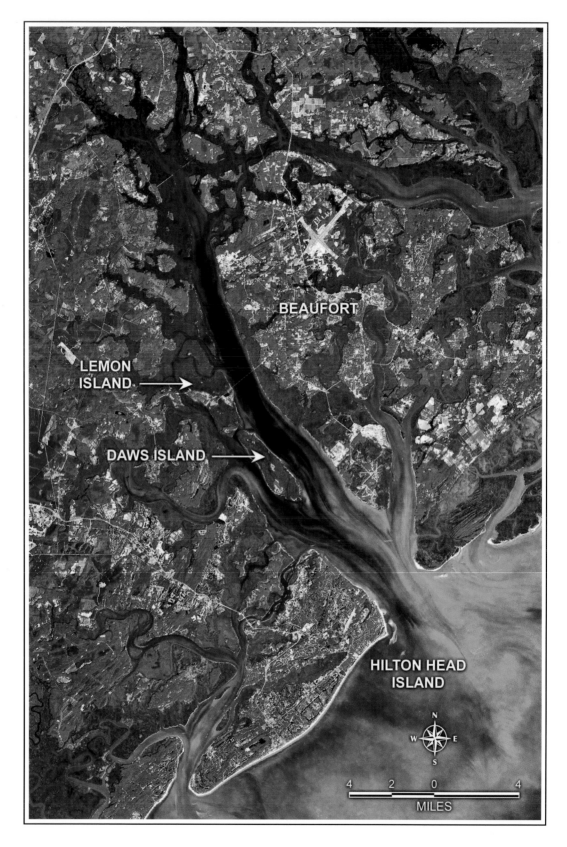

FIGURE 159. Vertical infrared image of Port Royal Sound acquired on 23 October 1999, courtesy of Earth Science Data Interface at the Global Land Cover Facility.

estuaries is a puzzle. One might speculate about fault lines, but we have not found any evidence to support such a correlation, as yet.

- There is a lot more open water in this estuarine complex than in the other estuaries in the Low Country.

- Therefore, very extensive marsh complexes, like those present in the upper reaches of St. Helena Sound (Figure 149), are not so common in this sound. This is most likely due to the fact that less suspended fine-grained sediments to build up tidal flats as a base for marshes is available in this sound, in comparison with St. Helena Sound, with its three significant black-water streams. The freshwater flow of the main tributary of the sound, the Coosawhatchie River, averages less than 200 cubic feet per second. This valley, having been carved so deep with little fresh water coming into it, just hasn't filled up with sediment yet.

- The color change to a purple tint of the waters in the uppermost parts of the sound show that at least some fresh water enters the system.

A brief historical note about the sound from the Wikipedia free encyclopedia informs us that in November, 1861, during the Second Civil War (of course), Union Commander Samuel F. DuPont disengaged the Confederate forts guarding the sound, and the area remained in Union hands for the rest of the war, becoming a major naval base. The Union Navy needed a deepwater harbor on the Southern coast if it were to maintain a year-round blockade of busy ports, such as Wilmington, Charleston, and Savannah. Because it was so deep and expansive, Port Royal Sound was probably the finest natural harbor on the southern coast at that time, thus it was ideally suited for the role. The day after the capture of Port Royal by the Union Commander DuPont, and, in effect, the establishment of a long Union reign in Port Royal Sound, General Robert E. Lee arrived in Savannah as the newly appointed commander of the South Atlantic coastal defenses. However, he was there

for only a few months, and by the following June, he was off to Virginia to become the leader of the Army of Northern Virginia. His first task there was to take on General George B. McClellan, President Linclon's favorite general (not!!), in a battle to save Richmond.

Places to Visit

Port Royal Sound is the most inaccessible subcompartment of the Low Country, unless you are touring with a watercraft. It is important to reiterate that the marsh system that borders the waters in the sound is mostly salt marsh, with only minimal brackish and freshwater marshes. The larger tributaries that enter the lower reaches of the sound, the Beaufort, Chechessee, and Colleton Rivers, are also dominated by salt marsh.

There is only one highway that crosses the sound, state route 170. As you cross the bridge heading west, cross the open water, and then cross the more minor channels on the west side, you will have several expansive marsh views. In the fall, you can see some red patches of *Salicornia* around the edge of the highlands in the marsh (right around the edge at the highest level). Also, broad expanses of dwarfed *Spartina alterniflora* are present. It is definitely a luxuriant marsh in this area, too bad there is no place to stop and look at it.

Route 170 crosses a Pleistocene upland called Lemon Island (Figure 159), which is located between the main body of the sound and the Chechessee River channel. We have never birded in this location, but the web page for Carolina Birds Club has the following comment about Lemon Island:

Lemon Island is relatively undeveloped and the sand flats and salt creeks along the highway harbor hundreds of shorebirds at high tide, such as plovers, oystercatchers, willets, dunlin, dowitchers, and whimbrel.

Reaching a spot to observe these birds may be

difficult, but there is a public boat ramp off state route 170 on Chechessee River at the Lemon Island Bridge, which could provide some access.

The Daws Island Heritage Preserve, a 1,885 acre island accessible only by water, is located a couple of miles south of the route 170 bridge. Manatees, dolphins, and sea turtles frequent the waters around the island, upon which ospreys and bald eagles nest.

This heritage preserve is a noteworthy archaeological site, where, during the past 40 years, archaeologists have identified over 25 sites associated with the first human occupants. A handful of sites date back to the Paleolithic (stone age), approximately 12,000 years ago, at which time sea level was considerably lower and Daws Island would have been more aptly named Daws "Hill." State archaeologist, Tommy Charles, remembers back to the late 1960s when beach combers and artifact collectors would roam the shoreline collecting thousands of beautifully flaked stone blades, informally known as "Daws Island points." Today, these ancient treasures of Daws Island are protected by state law. Primitive camping is welcomed, but it is limited and by permit only. Birding, camping, fishing, and kayaking are among the favorite activities of today's visitors to this pristine wildlife preserve (information on Daws Island provided by Tom Freeman).

HILTON HEAD HEADLAND

Morphology and History

The area under discussion in this subcompartment lies between the entrance to Port Royal Sound and the southwest end of Daufuskie Island (see satellite image in Figure 160). Two developed barrier islands, Hilton Head Island and Daufuskie Island, which are separated by a huge tidal inlet at the entrance to Calibogue Sound, are the centerpieces of the headland. The ebb-tidal delta of Calibogue Inlet merges in the offshore region with a large

shoal associated with the mouth of the Savannah River, a piedmont (red-water) river.

Similar to most of the large barrier islands on the Georgia coast, Hilton Head Island is composed of a seaward fringe of Holocene beach ridges that welded against a wide upland mass composed of Plesistocene sediments of the same age as those that underlie St. Helena Island on the northeast side of Port Royal Sound. As is typical of the low country, the sinking land has been flooded to the extent that both of the main islands are surrounded by large tidal channels in extensive salt marshes. Bull Island and Pinckney Island are Pleistocene remnants surrounded in the same fashion. Further landward, the Victoria Bluff Pleistocene upland is higher and older than the more seaward Pleistocene remnants. The historic town of Bluffton, on the banks of the May River, is located on the Victoria Bluff upland.

As is clear on the image in Figure 160, much of Hilton Head Island is developed and the same process is underway on Daufuskie, which has no roadway connection. Also, as you will notice if you visit the area, the Pleistocene upland areas inland of state road 170, and beyond, are undergoing intense development pressures at the present time.

This headland has played a prominent role in the human history of the South Carolina coast. Some of the most noteworthy events, summarized on the internet site, www.prismj.com/hiltonhead/hiltonheadhistory, are listed below:

- In 1521, a Spanish expedition entered Port Royal Sound and established a fortified town on what is now called Parris Island, the famous base of the U.S. Marine Corps. Over the next 20 years, Indian uprisings ultimately forced the abandonment of the fortified town, which was called Santa Elena.

- By 1860, there were 24 plantations in operation on Hilton Head Island, populated mostly by slaves and overseers who harvested the main crop, cotton. Because of the heat, as well as the island's low elevation (and mosquitos!!), the

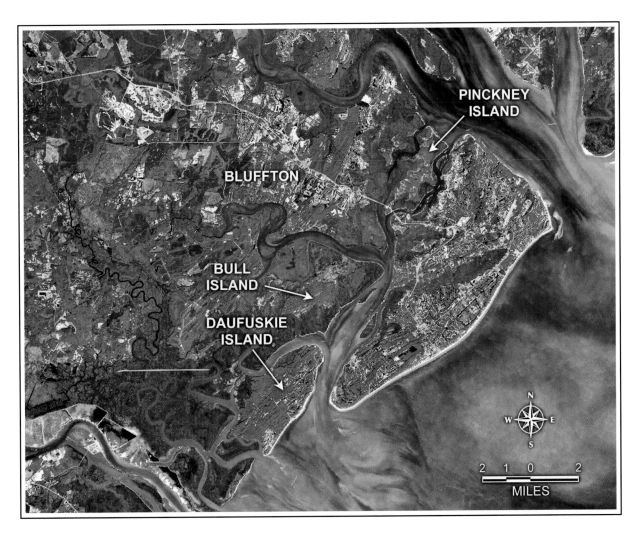

FIGURE 160. Vertical infrared image of Hilton Head headland acquired on 23 October 1999, courtesy of Earth Science Data Interface at the Global Land Cover Facility.

wealthy landowners spent little time on the island. This was true of many of the low country plantations. For example, in order to escape the mosquitos (and the diseases they vectored), many of the landowners further north went to the Edingsville "beach resort," which formerly existed south of Seabrook Island before it washed away.

- As noted earlier, this part of the world fell into Yankee hands early in the Second Civil War. When that happened and the Union Army moved onto Hilton Head, whatever island families were left, except the slaves presumably, evacuated their plantation homes. In 1862, with

the occupation of Union troops, missionaries, prisoners, and escaped slaves, Hilton Head's population swelled to over 40,000.

- During the 1940s and early 1950s, the island experienced a re-birth after a group of timbermen recognized the great potential in the island's tall straight pines. Popularly called sea pines, the trees produced lumber for a variety of uses. In 1956, Charles Fraser, son of one of the families who owned the island, realized that Hilton Head Island had more to offer than just timber. Armed with vision, energy, and investment dollars, he created a master plan for a resort community, Sea Pines

Plantation. That same year, a two-lane swing bridge was constructed, opening the island to the mainland. That bridge was replaced in 1982 with the current four-lane bridge. Charles Fraser was a legend among the coastal architects and developers of the 1970s, as we noted when the architects we were working with in Kuwait at the time were paying him their respects.

- The old bridge built in 1956 was named the James Byrnes Bridge, in honor of the preeminent South Carolina statesman of the 20th century. Born in Charleston in 1879, James Byrnes eventually opened a law office in Spartanburg. Among his expansive public service activities, he was elected to the U.S. House of Representatives, the U.S. Senate (for 10 years), and was an Associate Justice of the Supreme Court (appointed by President Roosevelt). He also served as Harry Truman's Secretary of State for two years, and concluded his public career by being Governor of South Carolina in the 1950s. A long-term advisor to both Presidents Roosevelt and Truman, he started out as a prominent New Deal Democrat, but like many southern Democrats of his time, he broke with the Democrats in the 1960s, supporting Nixon in 1960 and Goldwater in 1964. He died in Columbia in 1972 at age 92.
- In 1975, Hilton Head's population was 6,500, with 250,000 visitors to the island annually. In 1999, the population had swelled to over 30,000, with 2 million visitors annually.

Places to Visit

There are several key areas to visit on this protruding headland with the final destination of most tourists being Hilton Head Island itself. Die-hard naturalists may not find Hilton Head to their liking, but do not be too discouraged, because there is much to see on this headland. To get to the headland, one should continue south from the Port Royal Sound area on state road 170 for about 10 miles where you turn left (east) on US 278 and drive toward Hilton Head.

About 10 more miles along this road you will see a sign directing you to the Waddell Mariculture Center, where you should take a left on Sawmill Creek Road. About 1.4 miles along this road, you come to the **Victoria Bluff Heritage Preserve**, which is located on the high Pleistocene upland called Victoria Bluff mentioned earlier. The preserve grounds are open for public use from one hour before sunrise to one hour after sunset, unless otherwise posted or publicized. This preserve, which covers 1,111 acres, has fire breaks that serve as potential hiking trails through the **pine/saw palmetto flatwoods** community. These trees are dominated by longleaf, pond, and slash pines in the overstory and saw palmetto in the understory. In addition to saw palmetto, however, other evergreen species such as gallberry and fetterbush occur and form dense thickets in the understory.

The other plant community type located on the preserve is the **maritime forest**, which occurs adjacent to the saltwater creeks. It is dominated by live oak, cabbage palm, and slash pine. The understory in this habitat is somewhat sparse, with saw palmetto being the primary component.

We have never birded in this preserve, but it is said to be an excellent area for birding, particularly in the spring. Pine, yellow-throated, and yellow-rumped warblers, white-eyed vireos, summer tanagers, and flycatchers may be seen in a short time. White-tailed deer, fox, and gray squirrels may also be observed (information provided by the preserve's web page).

Continuing northeast along Sawmill Creek Road, you will reach the **Waddell Mariculture Center**, a research facility that has done pioneering work in the past two decades culturing a variety of species, including striped bass, white bass and their hybrids, two species of sturgeon, cobia, whiting, red drum, black drum, flounder, sea trout, tilapia, five species of marine shrimp, freshwater prawns, hard clams, bay scallops, and oysters. Perhaps their most noteworthy contribution that we are familiar with is the mariculture of **marine shrimp**. As noted on their web page:

"The Center has been the leader in intensification of pond production technology .. (for marine shrimp) .. By using the right combination of stocking density, feeding rate, aeration rate and water exchange, production as high as 29,249 kg/ha has been demonstrated at WMC. Some commercial farms in the area now routinely target production levels of 10,000 to 15,000 kg/ha."

Over the years, the authors of this book have had the pleasure of working with the scientists in this leading-edge research facility. Tours of the facility are provided several times a week. We think you would find such a tour to be exceptionally interesting.

After visiting these two areas on Sawmill Creek Road, return to US 278 and head southeast toward Hilton Head. After you cross the bridge over the Intracoastal Waterway take a left and proceed to the **Pinckney Island National Wildlife Refuge**, which is one in a chain of seven refuges comprising the Savannah Coastal Refuges network managed by the U.S. Fish and Wildlife Service. These refuges occur within a 100-mile stretch of coastline, with this refuge being the northernmost one. Pinckney Island Refuge is a historic cotton plantation.

We last visited the refuge on Thanksgiving weekend in 2006. We were there during a high spring tide, with the water right up against the road (check the tide tables; go at low tide!!). The individual islands in the refuge are surrounded by salt marsh. There are numerous informative signs scattered all along the walking trails, such as the one listing some of the most noteworthy birds – *black-crowned night heron, white ibis, wood stork, osprey, painted bunting, American oystercatcher, and snowy egret.*

The first part of the main trail from the parking area is along a dirt road, which starts out on a hammock with a climax maritime forest on it – live oak, sabal palmetto, some loblollies, saw palmetto, yaupon holly, and wax myrtles.

Another interesting sign describes salt-marsh plants – *glasswort (3 species), sea oxeye, and groundsel tree.*

"The marsh has developed in these harsh systems by developing strong leaf stems and roots, storing water in stems, expelling salt through leaves and roots, and growing shorter in more exposed areas."

You can ride a bike on this trail, which appears to be the only way you can get into the main part of the refuge. Another sign lists the creatures of the **salt pan** – *Atlantic ribbed mussel, marsh periwinkle, fiddler crabs, and oyster.* And another sign lists the birds of field and forest – *blue bird (year round); painted bunting (summer); yellow rumped warbler (winter); Carolina wren (year round); mourning dove, mocking bird, and cardinal (year round).*

A few hundred yards further along the trail you encounter **Wood Stork Pond** and a man-made pond called **Ibis Pond** (see photograph of one of these ponds in Figure 161). We saw the following birds at these ponds – common moorhen (gallinule), red-winged blackbird, wood duck, and coot. We also saw a number of yellow-bellied turtles, all at Ibis Pond. Wading birds we saw in the area included great blue heron, white ibis, juvenile white ibis, tricolored heron, green heron, little blue heron, black-crowned night heron, wood stork, snowy egret, and common egret. And, as a bonus, there are some gators in the ponds. In addition to the ponds, the refuge contains a diverse mixture of swampy woods, marshes, weedy fields, and pine plantations.

After leaving the refuge and continuing toward the ocean, you soon reach the **Coastal Discovery Museum** and **Visitor's Center** (just after passing the Hilton Head Island sign on the right). This visitor's

FIGURE 161. Pond on the Pinckney Island National Wildlife Refuge. Photograph taken on 25 November 2006.

center is chocked full of information related to the natural areas along the coast, such as:

"In South Carolina, 49% of the state's 120 breeding land birds are neotropical migrants, including barn swallow, painted bunting, northern parula warbler, yellow-breasted chat, and wood thrush."

In the center, there are either replicas or stuffed versions of the real thing of some of the most noteworthy extinct birds that formerly inhabited this area, namely the ivory-billed woodpecker, heath hen, passenger pigeon, and Carolina parakeet. This welcome center is a very informative place for a naturalist to visit, with a significant collection

of literature and maps to go along with the great displays.

Drive on toward the island and turn left at the next traffic light. Go down this road to the Green Shell Park, which is a little inconspicuous, on the left. From the parking area of the park, you can visit **Green Shell Enclosure Heritage Preserve**. A sign at the enclosure tells its story:

"Green shell enclosure, a low shell and earthen embankment is 2 to 6 feet in height, 30 feet wide at the base. During the Irene 1 period, 1,300 to 1,450 AD, this two-acre shell and earthen embankment was a fortified shelter village with defensive features, enclosures and palisades. Native American inhabitants of the Green

Shell Enclosure were farmers who lived in large villages. They used shell to make gourgets, masks, and beads. They also used shells for religious rituals. Inhabitants drank black drink, their ritual beverage, from whelk and conch shells. The black drink was made from the plant Ilex (Ilex vomitoria) commonly known as yaupon, which still grows in the preserve. The tree has small red berries, grows 5-20 feet, has small oval-shaped leaves with small ridges around the leaf edge. These villages not only used these items themselves but traded them far into the interior of the present United States.”

But 1,400 AD wasn't that long ago was it?

As a naturalist's stop, this site isn't too great, but it does emphasize the use of oyster shells and brings us a little closer to the people who originally inhabited this enticing part of the world.

Next stop, **Hilton Head Island**. We have done a little work on Hilton Head but no birding. If you are interested in birding on the island, you should buy a small book entitled *"Birder's Guide to Hilton Head Island, S. C. and the Low Country"* by the Hilton Head Audubon Society. They are available at the visitor's center. Surprisingly, there are a number of preserves and conservancies on the island that have nature trails and boardwalks. For example, a 605-acre preserve, **Sea Pines Forest Preserve**, is located in the Sea Pines Plantation (private with a $5 visitor's fee). The birding there is described as follows in the Birder's Guide:

"A few yards ahead is a 700-foot boardwalk over "Boggy Gut." This is one of those great little spots – all natural – quiet, except for red-winged blackbirds, moorhens, coots, kinglets (occasionally both species), blue-gray gnatcatchers, wrens, and perhaps a sora. Yes, he's there."

With regard to the beaches, as is true elsewhere in the state, all of the beach is public, from the low-tide line to the high water mark, which is hard to define, by the way. However, because of the intense development on this island, access to the beach from the road is limited to six places. Free public parking is available at only two of them, Fish Haul Park and Coligny Beach Park.

The beaches on the island are very flat and wide, being composed of fine-grained sand. Low intertidal bars develop in the intertidal zone between storms. According to NOAA, *"in 1990, a 6.6-mile section of the beach was renourished with close to 2.5 million cubic yards of sand. Two offshore sites, Gaskin Bank and Joiner Bank, served as sediment sources for the project. A pipeline dredge was set up to pump sediment onto the beach."* At least two nourishment projects preceded that one, and two more came later in 1997 and 2006. According to Tim Kana, over 10 million cubic yards of sand has been added to these beaches since 1969.

If you so desire, you can reach **Daufuskie Island** by regular ferry service. It is obviously more remote than Hilton Head and has some unique cultural aspects, but the coastal habitats and wildlife are not that different from what you can see at Hilton Head and the Pinckney Island National Wildlife Refuge. This island has experienced some beach erosion, with at least one nourishment project having been completed there in December 1998, when 1.4 million cubic yards of sand were added to the beach.

You may or may not know that Pat Conroy taught in an impoverished school on Daufuskie Island for a year in the 1960s and published an autobiographical account of his experiences in a book entitled *The Water Is Wide*. This island, which Conroy called Yamacraw in his book, is home to a culture and dialect with deep African roots, known as Gullah, or Geechee. In 1974, the movie *Conrack*, staring Jon Voight, was based on Conroy's book. Conrack was the way the students in his classes pronounced Conroy's name.

On the way south from Hilton Head, we recommend that you return via US 278, not that you have much choice unless you fly or swim or take a boat, and drive past the turn to the Waddell Mariculture Center. About 2 miles from that turn, take a left on state road 48 and visit the historic town of Bluffton. If you have spent much time on the burgeoning, overcrowded Hilton Head Island, this peaceful small town will probably be a welcome change. A Hilton Head/Bluffton Chamber of Commerce flyer tells the somewhat painful history of the town:

Bluffton's first small dwellings were constructed in the early 1800s on the May River bluffs. The layout of the town's streets in 1830 indicate that it had become a summer haven away from the mosquitoes and Yellow Fever problems at the rice plantations. The comforting southerly winds kept the mosquitoes at bay, making the sultry summer days bearable.

As the town grew, it became a commercial center for isolated plantations in the vicinity that received their goods from Savannah via the May River. Literally a hot bed for political rhetoric, in 1844 cries of secession were first given voice and debated here. With the Civil War raging and the eventual occupation of Hilton Head Island and Beaufort by Union forces, the town was mostly abandoned by residents and utilized as a base for Confederate pickets observing Union troop movements. The town was pillaged by Union forces on several excursions up the May River and eventually burned in June of 1863.

Although the overall destruction was severe, 15 houses and two churches survived.

You will see some stately old trees as you enter Bluffton. The reason the town was so pleasant in the summer time, relatively speaking, is that it was built on a high Pleistocene beach ridge (Victoria Bluff) at a point where the May River had cut a steep scarp in the upland. On a walking tour, you can visit the old houses and churches as well as get down to the shore at the base of the scarp where you can look out over the marsh. But be sure to keep an eye out for those Yankee rascals!

SAVANNAH RIVER DELTA

Morphology

The Savannah River delivers the second largest sediment load to the coast of any of the rivers in the Georgia Bight, second only to the combined output of the Santee and Pee Dee Rivers (Figure 162). Consequently, it has built the second largest river delta in the Bight. We are not aware of any detailed studies of the geology of the Savannah River delta similar to the one our group has conducted on the Santee/Pee Dee delta (discussed earlier). It is most likely a safe bet, however, that the two delta systems have a very similar origin and makeup.

Study the satellite image in Figure 163 to make the following observations:

- The abundance of the old rice ponds in the upper delta plain on the South Carolina side of the river, where the freshwater supplied by the Savannah River was available (now home of the Savannah National Wildlife Refuge).
- The city of Savannah on the south bank of the river, a very interesting historic city you may want to visit. Great architecture, and who wouldn't want to visit the setting for the book *Midnight in the Garden of Good and Evil* by John Berendt.
- A series of protruding barrier islands along the delta front to the south (Tybee, Little Tybee, and Wassaw), no doubt building out as a result of the input of sand from the river and the strong longshore southerly sand transport.
- Extensive salt and brackish marsh on the northeast side of the main channel of the river in the lower delta plain.
- Dredge spoil islands on the South Carolina side of the river.

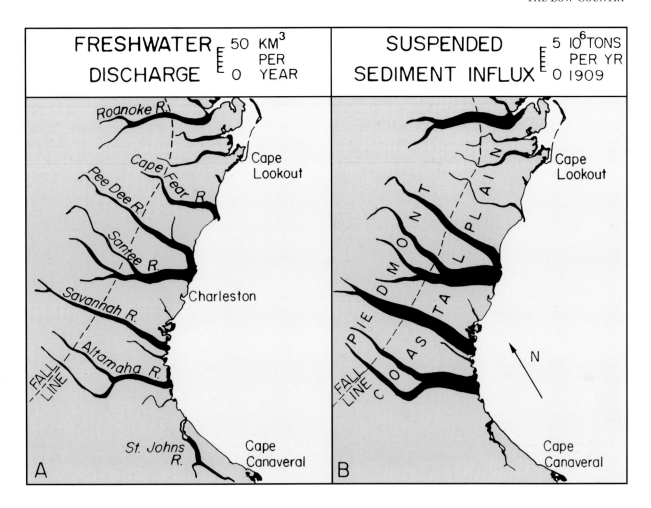

FIGURE 162. Water discharge (A) and suspended sediment influx (B) by the major rivers of the Georgia Bight. Freshwater discharge is based on 1931-1960 U.S. Geological Survey stream records, and suspended sediment discharge is based on data of Dole and Stabler (1909). Diagram is modified from Meade (1969) and Nichols and Biggs (1985).

Places to Visit

The **Savannah National Wildlife Refuge,** a visit to which no serious naturalist visitor to this coastline would want to miss, is on the **upper delta plain** of the Savannah River delta. To get there, if you are traveling south on I-95, take exit 5 off I-95 and go south on US 17. In the vicinity of Limehouse you pass through extensive pine forests, which provide some of the pulp for the paper-mill industry in the area. A couple of miles south of Limehouse, turn right on state road 170 and you are almost immediately on the delta plain and in the refuge. The valley wall is very subtle in this area. Going

west, you pass by a swamp that was logged not too long ago. After leaving the swamp, you will begin driving across restored rice ponds with abundant birds, such as the glossy ibis and red-tailed hawks, plus immeasurable numbers of ducks.

To enter the refuge proper, go almost to the river and then take a left to begin a 4-mile driving tour on levees that extend throughout the restored rice fields. The last time we did this drive was on Thanksgiving weekend in 2006. As we began the drive, we looked to the right where the port facilities for Savannah were conspicuous, as well as a paper mill.

Prescribed burning is one of the tools used

FIGURE 163. Vertical infrared image of Savannah River Delta acquired on 23 October 1999, courtesy of Earth Science Data Interface at the Global Land Cover Facility. Note the extensive dredged material disposal sites on the north side of the river.

by the refuge managers to eliminate the basis for exotic perennials, such as cattail, cutgrass species, and rattle bush. A sign by the drive explains the value of this process:

"Annual plants soon colonize the bare soil and yield seeds of higher value to wildlife. Burning removes tons of dry, dead vegetation that could fuel a destructive wild fire. The high value species are wild millet, giant foxtail, and fall panicum."

Just beyond mile 1, the road goes through an upland, island-like feature covered with trees, where another road sign informed us that:

*"This small island of trees was a slave community on a recessed plantation which bordered Laurel Hill Plantation called a **hammock**, a small area of high ground in a sea of rice fields. The round brick structure just ahead in the woods was a cistern that stored drinking water and perishable foods. …. Look for Chinese parasail trees with pale green trunks and 5-lobed leaves. The Asian trees were planted for shade on low country plantations. Notice periwinkle, a purple-flowered vine once cultivated as ground cover. Watch and listen for birds."*

If you follow that advice to watch for birds, you will not be disappointed, because they are everywhere!

At 1.5 miles, we spooked up a big flock of blue-winged teals (about 50) as well as some hooded mergansers and coots. We didn't see any gators on this drive, but we remembered a similar drive in the late winter several years ago when we came up on the largest gator we had ever seen while walking along one of the side trails. It was at least 12-13 feet long and covered with mud as it made its first "awakening" of the year.

At 3.5 miles, we stopped to admire what Hayes said was "the most ducks I have seen this side of Louisiana and Texas." There were thousands of

them, ring-billed ducks, widgeons, and so on. In addition to the multitude of ducks, we also saw flocks of ibises, abundant shore birds, including yellow legs, and several marsh hawks. In the early morning, this place has spectacular birding. If you are birding in the Low Country, you definitely do not want to miss the Savannah National Wildlife Refuge!

On the Georgia side of the river, you can visit the lower delta plain by driving down to Tybee Island. However, on the South Carolina side, it is only possible to reach the lower delta plain by driving the following route. Leave the refuge and drive east on state road 170 and turn right where this road joins US 17. Go south for about 3 miles and turn left on Alternate 170 where it branches off from US 17. Now go for about 3 more miles in a northeasterly direction until you reach SSR 92 (Shad Road), where you turn right and drive down to near the head of Wright River, where the road ends at a dirt ramp on a low hammock. This ramp, which would be a great place to launch a canoe or a kayak, is at the edge of natural brackish marsh, not an old rice field. This is an expansive *Juncus* marsh, which is host to an intensely meandering tidal channel as a result of the very flat slope of the lower delta plain. This channel system is shown in the aerial image given in Figure 163.

Thus ends our description of the Low Country compartment. With its remoteness, exceptional wetland complexes, and great birding, this is an outstanding place to either begin or end your tour of our Coast For All Seasons!

15 THE RIVERS AND SWAMPS

INTRODUCTION

As a naturalist visiting the coastline of the state of South Carolina, you would be wise to visit at least a few of the outstanding rivers and swamps that occur within the coastal plain. As shown in Figure 84A, there are three piedmont rivers in the state – the Pee Dee, Santee, and Savannah. On the other hand, there is a much larger number of coastal plain rivers. Three of them – the Wando, Cooper, and Ashley – empty into Charleston Harbor.

A spectacular engineering feat completed in 1941 diverted up to 90% of the flow of the Santee River (a piedmont river) into the Cooper River (a coastal plain river) by way of Lake Moultrie, a diversion canal, and through a hydroelectric plant at Pinopolis. Needless to say, this project has had a huge impact on the sedimentation rate within Charleston Harbor. This activity, which was completed in 1941, diverted an average of 15,000 cubic feet per second of fresh water (with its suspended fine-grained sediments) into the Cooper River, and subsequently, into the harbor itself. The result was an abrupt increase in shoaling within the harbor. As noted earlier, in 1986, the U.S. Army Corps of Engineers diverted over 50% of the flow back into the Santee River, partially mitigating this problem, presumably. Also, the mean monthly harbor surface salinity changed from 30.1 parts per thousand (ppt) to 16.8 ppt as a result of the diversion, and has again increased to 22.0 ppt since rediversion (Kjerfve and Magill, 1989). The complete engineering details of this monumental manipulation of a major river system is extremely interesting, but, unfortunately, is beyond the scope of this particular book.

PLACES TO VISIT

A pleasant driving tour you can take leaving from the city of Charleston would provide the opportunity to visit some of these rivers and swamps. To start the tour, drive south across the Ashley River bridge on US 17 and from there go north on state road 61 up the Ashley River Road, where you will be driving along the west side of the Ashley River. A few miles south of Magnolia Gardens, as indicated by the road signs, you pass by some subdivisions on the left (some are called plantations). Continue on to **Drayton Hall,** which neither of us has visited as yet! Their web page makes the following proclamation:

Welcome to Drayton Hall, a National Trust historic site in Charleston, South Carolina. Completed in 1742, the historic plantation house stands majestically on a 630-acre site and is one of the finest examples of Georgian-

Palladian architecture in America. Through seven generations of Drayton family ownership, the plantation house has remained in nearly original condition and offers an opportunity to experience history, to imagine the people—white and black—who lived and worked in a far different time.

From there, it is only one half mile more to **Magnolia Gardens**, a place we have visited several times. The birding is outstanding in these gardens, which are host to over 200 species over a year's time. The freshwater ponds and swamps have an abundance of wading birds, ducks, coots, and so on. The gardens and walkways, with a very photogenic white bridge, are some of the best in the area.

In about another 3.5 miles further along road 61, you come to **Middleton Place**, advertised to be "America's oldest landscaped garden." We have spent many pleasant evenings walking the fabulous landscaped trails on this property, it being a site where we had end-of-the-course banquets for some of our field seminars (after the demise of the Atlantic House Restaurant on Folly Island during hurricane *Hugo* in 1989). The restaurant is outstanding, as we are sure you have found many other restaurants in the Charleston area to be. The view over the Ashley River and the ancient rice ponds is uniquely pleasing, especially at dusk. If you saw the movie, *The Patriot*, you may remember a scene in which General Cornwallis was cooling his heals at Middleton Place, only to see a ship blown up out on the Ashley River. In any event, if you have a liking for landscape gardening, you don't want to miss this place.

Another short driving tour out of Charleston we recommend involves driving east across the Cooper River on I-526 and turning left on Clements Ferry Road as soon as you cross the bridge. About 6.5 miles further along, turn left on Cainhoy Road. Maybe all the non drivers should close their eyes as you pass a big industrial complex on the left, but somebody should keep their eyes open for a **red-cockaded woodpecker** *(Picoides borealis)* colony located a little further along on the right. There are a number of these colonies within the **Francis Marion National Forest**, the border of which is along the right side of this road. Some facts about this endangered woodpecker (from www.birding. about.com/library/weekly) follow:

- Very common in the southeastern United States in the mid-1800s.
- Now occupies about 1% of its original range.
- Its most distinguishing feature is a black cap and nape (the back of the neck) that encircle large white cheek patches.
- Makes its home in mature pine forests; more specifically, those with long-leaf pines averaging 80 to 120 years old and loblolly pines averaging 70 to 100 years old.
- The older pines favored by the red-cockaded woodpecker often suffer from a fungus called **red heart disease,** which attacks the center of the trunk, causing the inner wood to become soft.
- Each colony needs an average of 200 acres of old pine forest to support its foraging and nesting habitat needs.
- The nesting season lasts from April through June.
- Feeds mainly on beetles, ants, roaches, caterpillars, wood-boring insects, and spiders.

An easy way to recognize these colonies, besides the fact that most of the trees have a white ring painted around their trunks, is the lines of sap running down the side of the trees beneath the woodpecker's excavated living cavity. The source just cited explains the following reason for this:

Rat snakes are a primary predator of red-cockaded woodpeckers. Agile tree climbers, rat snakes eat woodpecker eggs and nestlings. But the red-cockaded woodpecker has an effective means of defense. It chips small holes (called resin wells)

260

in the bark of the cavity tree, above and below the cavity. From these wells, sap (resin) oozes down the trunk of the tree. When the snake comes in contact with the sap, the sap adheres to the scales on its underside. Even tiny amounts of resin inhibit the movement of the scales, preventing the snake from climbing higher.

Red-cockaded woodpeckers spend a significant amount of time and energy each day maintaining the flow of the resin wells. If the tree should die, or the damage from maintaining the wells becomes so great that the sap stops flowing, the woodpeckers will eventually abandon the cavity tree.

Continuing on the drive, the views are pretty mundane until you reach Blessing Lane. From there, you get into plantation land, passing by some bottomland hardwoods. There is a stately avenue of oaks at Middleburg Lane. We always enjoy this tour from Blessing Plantation north. This whole arm of the Cooper River is surrounded by historic plantations. We have visited Rice Hope Plantation, located in an enchanting setting on the banks of the Cooper River, which now has bed and breakfast living accommodations. You might want to check out the other plantations in the area on the Internet. At times, professional tours are offered for some of the plantation sites.

Once you reach the hamlet of Huger, turn right on Steed Creek Road and head toward Awendaw for a drive through the Francis Marion National Forest. About four miles in, you will drive through a pine savannah, with extensive sagebrush undergrowth throughout some scattered pines, which goes on for a few miles. In this middle area, you might want to explore some of the back roads, especially if you are birding. Over 300 species of migratory and non-migratory birds have been observed in the Forest. In addition to the already mentioned endangered red-cockaded woodpeckers, you may see swallow-tailed and Mississippi kites, bald eagles, wood storks, and numerous warblers and other song birds. At about

6.8 miles, you cross into Charleston County. *Hugo* hammered this area. There are no big pines left to speak of. Next is a stretch of water tupelo swamp that marks the boundary of the **Wambaw Swamp Wilderness,** 4,815 acres of swamp, which has no trails and little dry ground, and to add insult to injury, there is usually not enough water to float a canoe. Hmmm. Somebody said "this may be the least visited area in South Carolina." Good for the environment but not so good for the people?

Speaking of **swamps**, there are three outstanding, visitor-friendly swamp preserves that we highly recommend you visit – Cypress Gardens, Francis Beidler Forest (owned and operated by the National Audubon Society), and Congaree National Park. **Cypress Gardens** is located across the Cooper River from the plantations just discussed. Information and amenities of this locale, in addition to the usual good birding and photographic opportunities, include:

- The Gardens are open 7 days a week, from 9 am to 5 pm, except for Thanksgiving, Christmas Eve, Christmas Day, and New Years Day.
- Admission for adults is $10.
- Explore the interior of the swamp with flat bottom boats or walk on some winding trails around its edge.
- They also have a butterfly house, aquarium, reptile center, and aviary.

The other two recommended swamp sites are not nearly as close to Charleston as Cypress Gardens. To reach **Francis Beidler Forest** from Charleston, drive north on I-26 for about 25 miles and turn left at exit 187. From there, follow the signs for a few more miles to the Audubon Society preserve, which is located in Four Holes Swamp, a couple of miles to the east of I-26. This sanctuary, which contains over 15,000 acres (including 1,800 acres of virgin, old-growth forest), is open Tuesday through Sunday from 9 am until 5 pm, closed on Mondays and major holidays. Admission is $7 for adults.

We have visited this site many times in the past 20 years. The giant bald cypress trees (*Taxodium disticum*), with their massive cypress knees, are the most impressive we have seen. When you arrive, you should first go to the visitor's center, where a knowledgeable Audubon naturalist will be glad to answer any questions. After you leave the center, the first few hundred feet along the 1.75-mile long boardwalk pass through mostly upper bottomland hardwoods with abundant dwarf palmetto (*Sabel minor*) and a variety of standard upper bottomland hardwoods trees, including loblolly and spruce pines. Further along, you eventually encounter the lower bottomland hardwoods (the swamp), with very large bald cypress and water tupelo (*Nyssa aquatica*) trees. Like snakes? This is the place to see some giant cottonmouth water moccasins, as well as a variety of other reptiles. From the safety of the boardwalk, of course. The largest cottonmouth Hayes has ever seen was swimming toward him and Walter J. Sexton as they stood in knee-deep water trying to take a sediment core not far from the boardwalk. Fortunately, the big snake was not interested in them, and not afraid of them either; these are very aggressive snakes. With its microscopic freeboard, it skimmed almost effortlessly across the water surface toward the petrified geologists, missing them by less than 10 feet.

If you continue on the boardwalk from the edge of the swamp for several hundred yards, you will eventually come to a modest-sized lake, where, if you are quiet, you might see night herons, wood ducks, and possibly hear the two-syllabled cry of the red-shouldered hawk – *kee-yer, kee-yer*. See the photograph of this, the most common hawk in the swamp, in Figure 164.

One of the activities offered by the refuge is guided canoe/kayak floats, which are advertised as follows on their web site:

Paddle the still blackwater under towering 1,000-year-old cypress trees as an Audubon naturalist leads you deep into the heart of the Francis Beidler Forest in Four Holes Swamp. Take this opportunity to experience the largest remaining stand of virgin bald cypress and tupelo gum trees in the world in a way that most people do not.

The best trip we have ever had to Beidler was on one of the night-time guided walks along the boardwalk. The night we went was a crystal clear winter night, no moon. Throughout the swamp, you could hear the mournful calls of the famous "southern" owl, the barred owl – *Who cooks for you? Who cooks for you all?*

No red-blooded naturalist would want to miss a visit to this site. And how many times have we said that already?

To get to **Congaree National Park** from Charleston, drive north on I-26 for about 68 miles and turn right on US 601 at exit 145. From there, continue east on US 601 for about 19 miles, where you cross the bridge over the Congaree River (see photograph of the bridge in Figure 79). After you cross the four-mile-wide flood plain of the river, turn left on state road 48 and travel west for about 11 miles, and from there, follow the signs to the visitor's center for the park, which is open daily from 8:30 am to 5 pm, except Christmas day.

This preserve is another natural area we have visited many, many times, even before it first became a National Monument in 1976. It became a National Park in 2003. We always took the training seminars to this site (at least 150 times), and because we live so close, we have been birding and hiking in the park numerous additional times. Needless to say, we love going there, especially for the hiking trails and the birding in the spring.

There are many reasons any naturalist should go out of their way to visit this park, including:

- An informative visitor's center, which empha-sizes the largest intact tract of old-growth

FIGURE 164. Red-shouldered hawk, the most common hawk in the swamps of South Carolina.

bottomland hardwood forest in the United States.

- Approximately 90 tree species, half the number found in all of Europe, with many trees holding the state record for size.
- **Loblolly pines** (*Pinus taeda)* as tall as 169 feet, growing in a rare association with hardwood swamps.
- 22,200 acres of the floodplain of the Congaree River, including both upper and lower bottomland hardwoods, as well as several abandoned oxbow lakes.
- A canoe trail on a stream, Cedar Creek, that flows through the flood plain.
- There are 6 trails (offering 18 miles of hiking).
- An elevated boardwalk loop trail 2.4 miles long, which passes by some of the record trees as well as a beautiful oxbow lake (Weston Lake).

Figure 165 is a photograph taken from this boardwalk.

If you go in the spring, you will be able to hear the nesting warblers, such as the prothonatary, pine, and parula, singing throughout the forest. Red-shouldered hawks and barred owls also sometimes join in the chorus. All of the native woodpeckers are present, including a good number of pileateds. The real *guru* of the place is Ranger Fran Rametta, the architect and sometimes builder of the boardwalk, who has been there almost since the Monument was first opened. If you get a chance, get him to tell you the history of the park, and also try to persuade him to call up some **barred owls**. Fran is definitely worth the price of admission! Actually, the admission is free, but Fran should still be consulted if he is available.

FIGURE 165. View of the cypress/tupelo swamp on the flood plain of the Congaree River from the boardwalk in Congaree National Park. Photograph taken circa 2003.

You will be able to see some evidence of hurricane *Hugo* (1989), which blew down many trees in the park, including some record trees. We have taken the canoe trail a couple of times, which is worthwhile, but a little tough going in places, because of the numerous windfalls along the creek.

You will never be sorry you spent some time along the trails in this park.

16 THE FUTURE OF SOUTH CAROLINA'S BARRIER ISLANDS

A little over 10 years ago, Hayes published a paper on the barrier islands of the Georgia Bight (Hayes, 1994), which had a closing statement entitled "Future of the Barrier Islands." As the future of the barrier islands goes, so goes the South Carolina coast. The following discussion is a reproduction of the somewhat dire predictions made in that paper. Nothing much has happened in the past decade that would change these deductions.

As evidence continues to accumulate in favor of the global warming hypothesis (which is no longer in doubt, by the way), the prognosis for the future of the barrier islands of South Carolina is not good. Based on a mapping study to determine the potential impact of three levels of projected sea-level rise in the Charleston Harbor area, Kana et al. (1984) (a paper coauthored with the writers of this book) determined that Sullivans Island, a mixed-energy, prograding barrier island located just north of the harbor, would be converted to a narrow, landward-migrating barrier island around the year 2075 under the maximum sea-level rise scenario considered (7 feet in 80 years). This heavily populated island is now about 2,000 feet wide. It, and all the other prograding barrier islands in the state, would no doubt be eroded away under conditions of rapid sea-level rise. The new, landward-migrating (transgressive) barrier islands would be lodged against the interfluves between the expanded estuaries, and the waves would once again carve scarps into the Pleistocene core of Hilton Head Island, just as they did 4,500 years ago.

Furthermore, sand supply, the other half of the shoreline erosion/deposition equation, has been significantly impacted already in some localities by power dams and jetty systems. Also, under the rapid sea-level rise scenario, many of the important offshore sand sources, such as submerged delta lobes and major ebb-tidal deltas, could no longer contribute to the shoreline sand budget because of the increasing water depths.

Approximately one-half of the barrier islands described in this book are protected as national seashores, state and national wildlife refuges, and other environmentally oriented holdings. By most measures, these areas will not be a cause of serious concern, even if sea level does rise significantly and all the prograding islands are converted to landward-migrating washover terraces, except possibly for the loss of wetlands and overwintering habitat for waterfowl. However, wetland loss would probably be compensated for by the possible growth of wetlands in the expanded estuaries. The developed one-half of the coast is another story, and that is unfortunate, because there has been considerable progress made in the past 15 years of dealing with beach erosion problems on the developed shorelines of the state, principally in two areas:

- Fairly stringent beach protection and conservation laws have been passed. However, some of these types of legislation are not being seriously implemented, as yet, and some are being challenged legally, even to the level of the U.S. Supreme Court. Classic setbacks have been known to happen, such as the state of South Carolina's decision to basically abandon all regulation and encourage the construction of an "emergency dune" 9 feet high along 63 miles of the developed coast of the State a few days after hurricane *Hugo* (1989). Most of the sand was scraped from the lowest intertidal zone.

- There has been a move toward "soft engineering" solutions to erosion problems in developed areas, rather than building of hard structures, such as groins, seawalls, and rock revetments. An example of a soft solution is a project managed by our group at Captain Sams Inlet, where a huge volume of sand was made available to a severely eroding beach by relocating a small tidal inlet (covered in detail earlier in the discussion in the Barrier Islands compartment).

Under rapidly rising sea level, however, it seems likely that planners will take a more defensive posture and that these recent advances will be lost. Major defensive engineering structures will be built and many of the features we now know as barrier islands will become a series of isolated, heavily armored headlands. That is, unless the world's population gets serious about the problem of global warming.

RECOMMENDED ADDITIONAL READING

Abel, G. and Horan, J., 2001 (revised). Paddling South Carolina. Sandlapper Publishing Company, Inc., Orangeburg, SC, 160 pp.

Gives general information on paddling the rivers in the state, as well as specific facts, including places to put in and take out, for the 30 most floatable rivers.

Ballantine, T., 1991. Tideland Treasure. University of South Carolina Press. 218 pp.

According to the publisher, this book "is an illustrated guide to the beaches and marshes of the Eastern United States coast. .... Lavished with true-to-life illustrations ..(over 400 of them).. and hand-written text, the book portrays the nature of the sea, beach, salt marsh, plants, and animals of the area in everyday language."

Berendt, J., 1997, Midnight in the Garden of Good and Evil. Vintage Books, New York, 400 pp.

Good book to read to get you in the right frame of mind for visiting the historic districts of Charleston and Savannah (the setting for the book). The makers of a movie by the same name went out of their way to avoid doing justice to this book.

Blagden, T., Jr. (photographer), with a text by J. Lareau and R. Porcher, 2003, Lowcountry – The Natural Landscape. Legacy Publications, Greensboro, N.C., 104 pp.

Said by John Henry Dick (coastal artist) to be "the first exhibit-format volume to celebrate the essence of wilderness embodied in the rivers, forests, swamps, marshes, and barrier islands of South Carolina." Outstanding photographs and brief descriptions of many of the coastal habitats discussed in this book.

Davis, R.A. Jr., and FizGerald, D.M., 2004. Beaches and Coasts. Blackwell Science Limited, Oxford, U.K., 419 pp.

The best modern textbook for giving the novice an introduction to the physical processes of the coastal zone.

Gore, A., 2006. An Inconvenient Truth. Rodale Books, Emmaus, PA, 325 pp.

Well-illustrated discussion of most of the issues related to global warming, such as melting of glaciers and icefields and increased CO_2 in the atmosphere.

Hatcher, R.D., Jr., 2005. Southern and Central Appalachians. In: Selley, R.C., Cocks, R., and Plimer, I. (eds), Encyclopedia of Geology. London, Elsevier Ltd., pp. 72-81.

Beautifully illustrated and thorough treatment of the origin of the Appalachian Mountains.

Hilton Head Audubon Society (Graham C. Dugas, Jr.), 1997. Birder's Guide to Hilton Head Island, SC and the Low Country. 56 pp.

Good source of information on places to visit to find birds on the island, as well as other parts of the Low Country of South Carolina.

Kana, T.W., 2003. Coastal Erosion and Solutions: A Primer. Coastal Science and Engineering LLC., Columbia, SC, 25 pp.

Concise introduction to the problem, with emphasis on erosion on the South Carolina coast. Many well-illustrated examples of the processes and products of coastal erosion in the state.

Kana, T.W., 1990. Conserving South Carolina Beaches Through the 1990s: A Case for Beach Nourishment. S.C. Coastal Council and NOAA, 30 pp.

Introduction to the topic of beach nourishment, with examples from along the South Carolina coast in the 80s and a look at possible projects in the 90s.

Lennon, G., Neal, W.J., Bush, D.M., Pilkey, O.H., Stutz, M., and Bullock, J., 1996. Living with the South Carolina Coast: Duke University Press, 241 p.

One in a series of books published by Orrin Pilkey and associates on the theme of "how to live with the coast." In addition to an excellent bibliography and detailed discussion of the effects of hurricane *Hugo*, there are blow-by-blow discussions of the different segments of the coast with relation to risk zones, such as where to expect flooding during storms, and so on.

Meyer, P., 1991. Nature Guide to the Carolina Coast: Common Birds, Crabs, Shells, Fish and Other Entities of the Coastal Environment. Avian-Cetacian Press.

General treatment of the subject with some details, including photographs, of the most common shells, fish, and birds, among other inhabitants of the shore.

Murphy, C.H., 1995. Carolina Rocks! Sandlapper Publishing Co., Inc., Orangeburg, SC, 261 pp.

Good introduction to all aspects of the geology of South Carolina, with detailed discussion of certain topics, such as the Charleston earthquake of 1886 and the Carolina Bays.

Pilkey, O.H., and Fraser, M.E., 2003. A Celebration of the World's Barrier Islands. Columbia University Press, New York, 309 pp.

Beautifully illustrated coverage of barrier islands in general, with several references to those in South Carolina. Excellent introduction for one just getting interested in barrier islands.

Potter, E.F., Parnell, J.E., and Teulings, R.P., 1980. Birds of the Carolinas. University North Carolina Press, Chapel Hill, 406 pp.

Well-illustrated coverage, including a number of photographs, of all the bird species that occur within the two states. Discussion of each species typically includes range, nesting habits, and feeding habits.

Rubillo, T., 2006, Hurricane Destruction in South Carolina – Hell and High Water. History Press, Charleston, SC, 147 pp.

Discusses all the major hurricanes to hit the South Carolina coast since the beginning of the recorded history of the area, ending with a discourse on the infamous *Hugo*. Very informative. Great personalized accounts from times before detailed records were kept, as well as good details on the later storms.

Rutledge, A., 1960, 1966 (revised edition). Deep River, The Complete Poems of Arichibald Rutledge. R. L. Bryan Co., Columbia, SC, 635 pp.

This volume includes all of his poems, but two of the sections, *Heart of the Wildwood* and *Deep River*, have many poems based on his experiences on the South Carolina coast.

Sibley, D.A., 2003. The Sibley Field Guide to Birds of Eastern North America. A.F. Knopf, New York, 347, pp.

This is the field guide we use. Outstanding coverage with beautiful artwork.

Sprint, A., Jr., and Chamberlain, E. B., 1949, 1977 (second printing). South Carolina Bird Life. University of South Carolina Press, Columbia, 658 pp.

The classic reference on the birds of the state. Not a bird guide, but a great source for such details as when a certain species was first described in the state, details on their feeding habits, their range, and so on. Some photographs and a number of paintings by famous wildlife artists, such as Roger Tory Petterson and John Henry Dick, are also included. When we have a question about some attribute of one of the species we see only on occasion, this is the first place we look for an answer.

Tiner, R.W. and Rorer, A., 1993. Field Guide to Coastal Wetland Plants of the Southeastern United States. University of Massachusetts Press, 336 pp.

Covers the wide variety of plants present in the coastal zone of South Carolina.

REFERENCES CITED

Abel, G. and Horan, J., 2001 (revised). Paddling South Carolina. Sandlapper Publishing Company, Inc., Orangeburg, SC, 160 pp.

Allen, J.R.L., 1968. Current Ripples – Their Patterns in Relation to Water and Sediment Motion. North-Holland Publishing Company, Amsterdam, Holland, 433 pp.

Baldwin, W.E., Morton, R.A., Schwab, W.C., Gayes, P.T., and Driscoll, N.W., 2004. Maps showing the stratigraphic framework of South Carolina's Long Bay from Little River to Winyah Bay. U.S. Geological Survey, Open-File Report, OF 2004-1013.

Baldwin, W.E., Morton, R.A., Putney, T.R., Katuna, M.P., Harris, M.S., Gayes, P.T., Driscoll, N.W., Denny, J.F., and Schwab, W.C., 2006. Migration of the Pee Dee River System Inferred from Ancestral Paleochannels Underlying the South Carolina Grand Strand and Long Bay Inner Shelf. Geological Society of America Bulletin, v. 118, no. 5/6, pp. 533-549.

Barry, J.M., 1980. Natural Vegetation of South Carolina. University of South Carolina Press, Columbia, SC, 214 pp.

Basan, P.B. and Frey, R.W., 1977. Actual-palaeontology and Neoichnology of Salt Marshes Near Sapelo Island, Georgia. In: Crimes T.P, and Harper, J.C. (eds.), Trace fossils 2. Geographical Journal, Special Issue 9, pp. 41-70.

Bascom, W.H., 1954. Characteristics of Natural Beaches. Proceedings, 4th Conference on Coastal Engineering, pp. 163-180.

Berendt, J., 1997. Midnight in the Garden of Good and Evil. Vintage Books, New York, NY, 400 pp.

Biggs, R.B., 1978. Coastal Bays. In: Davis, R. A., Jr. (ed.), Coastal Sedimentary Environments. Springer Verlag, New York, pp. 69-99.

Bird, E.C.F., 1996. Beach Management. Coastal Morphology and Research. John Wiley, Hoboken, NJ, 292 pp.

Boothroyd, J.C., 1969. Hydraulic Conditions Controlling the Formation of Estuarine Bedforms. In: Hayes, M.O. (ed), Coastal Environments – NE Massachusetts and New Hampshire: Field Trip Guidebook for Eastern Section of SEPM, May 9-11, 1969. pp. 417-427.

Bowen, A.J., and Inman, D.L., 1966. Budget of Littoral Sands in the Vicinity of Point Arguello, California. Coastal Engineering Research Center, Vicksburg, MS, Tech. Memo. no. 19, 56 pp.

Brown, D., 1970. Bury My Heart at Wounded Knee: An Indian History of the American West. H. Holt and Co., New York, 487 pp.

Cajete, G.A., 1995, Ensoulment of nature. *In:* Hirschfelder, A.B. (ed), Native Heritage: Personal Accounts by American Indians, 1790 to the Present. Macmillan Publishing Company, New York, NY, 298 pp.

Clark, J.R. and Benforado, J. (eds), 1981. Wetlands of Bottomland Hardwood Forests. Elsevier, Amsterdam, 401 pp.

Coleman, J.M., and Wright, L.D., 1975. Modern River Deltas: Variability of Processes and Sand Bodies. *In:* Deltas, M.L. Broussard (ed), Houston Geological Society, Houston, TX. 2nd ed., pp. 99-150.

Colquhoun, D.J., et al., 1983. Surface and Subsurface Stratigraphy, Structure and Aquifers of the South Carolina Coastal Plain. Report for S.C. Dept. of Health and Environmental Control, Dept. Geology, University of South Carolina, Columbia, SC, 78 pp.

Colquhoun, D.J. and Brooks, M.J., 1986. New Evidence from the Southeastern U.S. for Eustatic Components in the Late Holocene Sea Levels. Geoarchaeology, v. 1, pp. 275-291.

Comet, P.A., 1996. Geological Reasoning: Geology as an Interpretive and Historical Science; Discussion. Geological Society of America Bulletin, v. 108, no. 11, pp. 1508-1510.

Concerned Coastal Geologists, 1981. Saving the American Beach. Skidaway Institute of Oceanography. Conference, Savannah, GA, March 1981, 12 pp.

Conner, W.C., Kraft, R.H., and Harris, D.L., 1957. Empirical Methods for Forecasting the Maximum Storm Tide Due to Hurricanes and Other Tropical Storms. Monthly Weather Review, v. 85, pp. 113-166.

Conroy. P., 1972. The Water is Wide. Published in paperback by Bantam, New York in 1995.

Conroy, P., 1986. The Prince of Tides. Houghton Mifflin Company, Boston, MA, 576 pp.

Cote, R.N., 2006. City of Heroes. Corinthian Books, Mt. Pleasant, SC, 542 pp.

Curray, J.R., 1964. Transgressions and Regressions. *In:* Papers in Marine Geology – Shepard Commemorative Volume, Miller, R.L. (ed), MacMillian and Co., New York, NY, pp. 175-203.

Dean, R.G. and Walton, T.C., 1975. Sediment Transport Processes in the Vicinity of Inlets with Special Reference to Sand Trapping. *In:* L.E. Cronin (ed), Estuarine Research, Geological and Engineering. Academic Press, New York, v. 2, pp. 129-150.

DePratter, C.B. and Howard, J.D., 1980. Indian Occupation and Geologic History of the Georgia Coast: A 5,000 Year Summary. *In:* Howard, J.D., DePratter, C.B., and Frey, R.W. (eds), Excursions in Southeastern Geology: Archaeology of the Georgia Coast. Georgia Dept. Natural Resources, Atlanta, GA, pp. 1-65.

Dietz, R.S., and Holden, J.C., 1970. Reconstruction of Pangaea: Breakup and Dispersion of Continents, Permian to Present. Journal of Geophysical Research, v. 75, pp, 4939-4956.

Dole, R.B. and Stabler, H., 1909. Denudation. U.S. Geological Survey Water Supply Paper 234, pp. 78-93.

Donovan-Ealy, P., Gayes, P.T., Nelson, D.D., Van Dolah, R. R., and Maier, P., 1993. Development of a Database of Geologic Resources and Critical Habitats on the South Carolina Continental Shelf. *In:* Proceedings of the Geological Information Society, v. 24, pp. 173-178.

Dugas, Jr., G.C., 1997. Birder's Guide to Hilton Head Island, SC and the Low Country. Hilton Head Audubon Society, Hilton Head, SC, 56 pp.

DuMars, A.J., 2007. Shoreline Process and Response: Northeastern Folly Island, South Carolina. Final Rept., Charleston County Parks and Recreation Commission, Charleston, SC, 35 pp.

Dunn, G.E. and Miller, B.I., 1960. Atlantic Hurricanes. Louisiana State University Press, Baton Rouge, LA,

326 pp.

Eckert, A.W., 1993. A Sorrow in Our Hearts, The Life of Tecumseh. Bantam Books, New York, NY, 1088 pp.

Finkl, C.W. and Pilkey, O.H. (eds), 1991. Impacts of Hurricane *Hugo*: September 10-22, 1989. Journal of Coastal Research, Special Issue 8, 356 pp.

Fischer, J.J., 1967. Origin of Barrier Island Chain Shoreline, Middle Atlantic States. Geological Society of America, Special Paper 115, pp. 66-67.

FitzGerald, D.M., 1977. Hydraulics, Morphology, and Sediment Transport at Price Inlet, South Carolina. Ph.D. Thesis, Dept. Geology, University of South Carolina, Columbia, SC, 84 pp.

FitzGerald, D.M. and Hayes, M.O., 1980. Tidal Inlet Effects on Barrier Island Management. Coastal Zone '80, American Society of Civil Engineering, New York, NY, pp. 2355-2379.

Freeman, J.C., Baer, L., and Jung, G.H., 1957. The Bathystropic Storm Tide. Journal of Marine Research, v. 16 pp. 12-22.

Frey, R.W. and Howard, J.D., 1969. A Profile of Biogenic Sedimentary Structures in a Holocene Barrier Island-Salt Marsh Complex. Trans. Gulf Coast Association of Geological Societies, v. 19, pp. 427-444.

Galloway, W.E., 1975. Process Framework for Describing the Morphology and Stratigraphic Evolution of Deltaic Depositional Systems. *In:* M.L. Broussard (ed), Deltas, 2nd Edition, Houston Geological Society, Houston, TX, pp. 87-98.

Gayes, P.T., 1991. Post-hurricane *Hugo* Nearshore Side Scan Sonar Survey: Myrtle Beach to Folly Beach, South Carolina. *In:* Finkl, C.W. and Pilkey, O.H. (eds), Impacts of Hurricane *Hugo*: September 10-22, 1989. Journal of Coastal Research, Special Issue 8, pp. 95-111.

Gayes, P.T., Schwab, W.C., Driscoll, N.W., Morton, R.A., Baldwin, W.E., Denny, J.J., Wright, E.E., Harris, M.S., Katuna, M.P. and Putney, T.R., 2003, Sediment Dispersal Pathways and Conceptual Sediment Budget for a Sediment Starved Embayment: Long Bay. *In:* Davis, R.A, Jr., Sallenger, A., and Howard, P., (eds.), Coastal Sediments '03, American Society of Civil Engineers, St. Petersburg, Florida.

Gilbert, G.K., 1889. The Topographic Features of Lake Shores. U.S. Geological Survey 5th Annual Report, Washington, DC, pp. 69-123.

Glaeser, J.D., 1978. Global Distribution of Barrier Islands in Terms of Tectonic Setting. Journal of Geology, v. 86, pp. 283-298.

Glen, W. 1975. Continental Drift and Plate Tectonics: C.E. Merrill Publishing Company, Columbus, OH, 188 pp.

Gore, A., 2006. An Inconvenient Truth. Rodale Books, Emmaus, PA, 325 pp.

Griffin, M.M., 1981. Feldspar Distribution in River, Delta, and Barrier Sands of Central South Carolina. M.S. Thesis, Dept. of Geology, University of South Carolina, Columbia, SC, 122 pp.

Hands, E.B., 1977. Implications of Submergence for Coastal Engineers. Coastal Sediments '77, Proceedings, American Society of Civil Engineers, New York, NY, pp. 149-166.

Harris, D.L., 1963. Characteristics of the Hurricane Storm Surge. U.S. Weather Bureau Technical Paper No. 48, 139 pp.

Harris, M.S., Gayes, P.T., Kindinger, J.L., Flocks, J.G., Krantz, D.E., and Donovan, P., 2005. Quaternary Geomorphology and Modern Coastal Development in Response to an Inherent Geologic Framework: An Example from Charleston, South Carolina. Journal of Coastal Research, v. 21, No. 1, pp. 49-64.

Hatcher, R.D., Jr., 2005. Southern and Central Appalachians. *In:* Selley, R.C., Cocks, R., and Plimer, I. (eds),

Encyclopedia of Geology. Elsevier Ltd., London, pp. 72-81.

Hayes, M.O., 1965. Sedimentation on a Semiarid, Wave-dominated Coast (south Texas); with Emphasis on Hurricane Effects. Ph.D. Thesis, Dept. of Geology, University of Texas, Austin, TX, 350 pp.

Hayes, M.O., 1967. Hurricanes as Geological Agents: Case Studies of Hurricanes *Carla* (1961) and *Cindy* (1963). University of Texas, Austin, Bureau of Economic Geology Report 61, 56 pp.

Hayes, M.O., 1969. (ed), Coastal Environments – NE, Massachusetts and New Hampshire: Field Trip Guidebook for Eastern Section of SEPM, May 9-11. 462 pp.

Hayes, M.O., 1976. Lecture Notes. *In:* Hayes, M.O. and Kana, T.W. (eds), Terrigenous Clastic Depositional Environments. Tech Rep 11-CRD, Geology Dept., University of South Carolina, Columbia, SC, pp. I-1 – I-131.

Hayes, M.O., 1977. Development of Kiawah Island, South Carolina. Coastal Sediments '77, American Society of Civil Engineers, Charleston, SC, pp. 828-847.

Hayes, M.O., 1979. Barrier Island Morphology as a Function of Tidal and Wave Regime. *In:* Leatherman, S. (ed), Barrier Islands, from the Gulf of St. Lawrence to the Gulf of Mexico. Academic Press, New York, NY, pp. 1-27.

Hayes, M.O., 1980. General Morphology and Sediment Patterns in Tidal Inlets. Sedimentary Geology, v. 26, pp. 139-156.

Hayes, M.O., 1985. Beach Erosion. *In:* Clark, J.R. (ed), Coastal Resources Management: Development Case Studies. National Park Service and U.S. Agency for International Development, Washington, DC, pp. 67-201.

Hayes, M.O., 1994. Georgia Bight. Chapter 7 *in* R.A. Davis, Jr. (ed), Geology of the Holocene Barrier Island System. Springer-Verlag, Berlin, pp. 233-304.

Hayes, M.O., 1999. Black Tides. University of Texas Press, Austin, TX, 287 pp.

Hayes, M.O. and Boothroyd, J.C., 1969. Storms as Modifying Agents in the Coastal Environment. *In:* M.O. Hayes (ed), Coastal Environments – NE Massachusetts and New Hampshire: Field Trip Guidebook for Eastern Section of SEPM, May 9-11. pp. 290-315.

Hayes, M.O., Hulmes, L.J., and Wilson, S.J., 1974. Importance of Tidal Deltas in Erosional and Depositional History of Barrier Islands (Abstract). Geological Society of America, Abstracts with Programs, 1974 Ann. Mtg.

Hayes, M.O., Wilson, S.J., FitzGerald, D.M., Hulmes, L.J., and Hubbard, D.K., 1975. Coastal Processes and Geomorphology. *In:* Environmental Inventory of Kiawah Island. Environmental Research Center, Inc., Columbia, SC, 165 pp.

Hayes, M.O. and Kana, T.W. (eds), 1976. Terrigenous Clastic Depositional Environments. Tech Rep 11-CRD, Geology Dept., University of South Carolina, Columbia, SC, 315 pp.

Hayes, M.O., FitzGerald, D.M., Hulmes, L.J., and Wilson, S.J., 1976. Geomorphology of Kiawah Island, South Carolina. *In:* Hayes, M.O. and Kana, T.W. (eds), Terrigenous Clastic Depositional Environments. Technical Report 11-CRD, Geology Dept., University of South Carolina, Columbia, SC, pp. II-80 – II-100.

Hayes, M.O., Kana, T.W., and Barwis, J.H., 1980. Soft Designs for Coastal Protection at Seabrook Island, SC. Proc. 17th Conference on Coastal Engineering, American Society of Civil Engineers, New York, Chapter 56, pp. 897-912.

Hayes, M.O., and Sexton, W.J., 1983. Prognosis of Future Shoreline Changes on Botany Bay Island, South

Carolina. Report for Cubit Engineering, Ltd., RPI Rep 83-4, Research Planning, Inc., Columbia, SC, 15 pp.

Hayes, M.O., Sexton, W.J., and Sipple, K.N., 1994. Fluid-bearing Capacity of Strandline Deposits-Implication for Hydrocarbon Exploration (Abstract). American Association of Petroleum Geologists Bulletin, v. 68, p. 485.

Hayes, M.O. and Sexton, W.J., 1989. Field Trip Guidebook T371: Modern Clastic Depositional Environments, South Carolina. 29th International Geological Congress, American Geophysical Union, Washington, DC, 85 pp.

Hine, A.C., 1979. Mechanisms of Berm Development and Resulting Beach Growth Along a Barrier Spit Complex. Sedimentology, v. 26, pp. 333-351.

Hosier, P.E., 1975. Dunes and Marsh Vegetation. In: Environmental Inventory of Kiawah Island. Environmental Research Center, Inc., Columbia, SC, pp. D-1 – D-96.

Howard, J.D. and Dorjes, J., 1972. Animal-sediment Relationships in Two Beach-related Tidal Flats, Sapelo Island, Georgia. Journal of Sedimentary Petrology, v. 42, pp. 608-623.

Hoyt, J.H. and Weimer, R.J., 1963. Comparison of Modern and Ancient Beaches. American Association of Petroleum Geologists Bulletin, v. 47, pp. 5529-5531.

Hubbard, D.K., 1977. Variations in Tidal Inlet Processes and Morphology in the Georgia Embayment. Technical Report, 14-CRD, Dept. Geology, University of South Carolina, Columbia, SC, 79 pp.

Imperato, D.P., Sexton, W.J. and Hayes, M.O., 1983. A Mechanism for Beach Ridge Formation: Abstract, Annual Meeting, Geological Society of America, Indianapolis, IN.

Imperato, D.P., Sexton, W.J., and Hayes, M.O., 1988. Stratigraphy and Sediment Characteristics of a Mesotidal Ebb-tidal Delta, North Edisto Inlet, South Carolina. Journal of Sedimentary Petrology, v. 58, pp. 950-958.

Inman, D.L. and Nordstrom, C.E., 1971. On the Tectonic and Morphologic Classification of Coasts. Journal of Geology, v. 79, pp. 1-21.

Ivester, A.H., Brooks, M.J., and Taylor, B.E., 2007. Sedimentology and Ages of Carolina Bay Sand Rims (Abstract). Geological Society of America, Southeastern Section–56th Annual Meeting, Savannah, GA, March 2007, pp. 29–30.

Kaczorowski, R.T., 1976. Origins of the Carolina Bays. University of South Carolina Technical Report, No. I I-CRD, Pt. 2, Columbis, SC, pp. 16-36.

Kana, T.W., 1976. Coastal Processes and Sediment Transport at Price and North Inlets, South Carolina. M.S. Thesis, Dept. of Geology, University of South Carolina, Columbia, SC, 144 pp.

Kana, T.W., 1977. Suspended Sediment Transport of Price Inlet, South Carolina. Coastal Sediments '77, American Society of Civil Engineers, pp. 366-382.

Kana, T.W., 1979. Suspended Sediments in Breaking Waves. Ph.D. Thesis, Dept. of Geology, University of South Carolina, Columbia, SC, 153 pp.

Kana, T.W., 1988. Beach Erosion in South Carolina. Report for South Carolina Sea Grant Consortium, Columbia, SC, 55 pp.

Kana, T.W., 1989. Erosion and Beach Restoration at Seabrook Island, South Carolina. Shore and Beach, v. 57, pp. 3-18.

Kana, T.W., 1990. Conserving South Carolina Beaches Through the 1990s: a Case for Beach Nourishment. S.C. Coastal Council and NOAA, Charleston, SC, 30 pp.

Kana, T.W., 2006. Myrtle Beach Restoration: A Success Story? v. 52, No. 4, Business & Economic Review, July-September 2006.

Kana, T.W., Michel, J., Hayes, M.O., and Jensen, J.R. 1984. The Physical Impact of Sea-level Rise in the Area of Charleston, SC. *In:* Barth, M.C. and Titus J.G. (eds), Greenhouse Effect and Sea Level Rise. Van Nostrand Reinhold Co., New York, NY, pp. 105-150.

Kana, T.W. and Mason, J.E., 1988. Evolution of an Ebb-tidal Delta after Inlet Relocation. *In:* Aubrey, D.G. and Weishar, L. (eds), Hydrodynamics and Sediment Dynamics of Tidal Inlets. Springer-Verlag, New York, NY, pp. 382-411.

Kana, T.W., Hayter, E.J., and Work, P.A., 1999. Mesoscale Sediment Transport at Southeastern U.S. Tidal Inlets: Conceptual Model Applicable to Mixed Energy Settings. Journal of Coastal Research, v. 15, pp. 303-313.

Kana, T.W., and Gaudiano, D.J., 2001. Regional Beach Volume Changes for the Central South Carolina Coast. Technical Report, Dept. of Geological Sciences, University of South Carolina, Columbia, SC, 124 pp.

Kana, T.W., and McKee, P., 2003. Relocation of Captain Sams Inlet – 20 Years Later: Coastal Sediments '03, American Society of Civil Engineers, New York, NY, 12 pp.

Kana, T.W., White, T.E., and McKee, P., 2004, Management and Engineering Guidelines for Groin Rehabilitation. Journal of Coastal Research, v. 33, pp. 57-82.

Kjerfve, B. and Magill, K.E., 1989. Hydrological Changes in Charleston Harbor. Coastal Zone '89, v. 3, Proc. 6th Symposium of Coastal and Ocean Management, American Society of Civil Engineers, New York, NY, pp. 2640-2649.

Kjerfve, B. and Magill, K.E., 1990. Salinity Changes in Charleston Harbor 1922-1987. J. Waterway Port Coastal Ocean Engineering, v. 116, pp. 153-168.

Knoth, J.S. and Nummedal, D., 1977. Longshore Sediment Transport Using Fluorescent Tracer. Coastal Sediments '77, American Society of Civil Engineers, New York, NY, pp. 383-398.

Komar, P.D., 1976. Beach Processes and Sedimentation. Prentice-Hall, Inc., Englewood Cliffs, N.J., 429 pp.

Larson, E., 1999. Isaac's Storm: A Man, a Time, and the Deadliest Hurricanes in History. Crown, New York, NY, 336 pp.

Lawson, J., 1709. A New Voyage to Carolina. 1976 edited version by H.T. Lefler, UNC Press, Chapel Hill, NC, 305 pp.

LeGrand, H.E., 1961. Summary of Geology of Atlantic Coastal Plain. American Association of Petroleum Geologists Bulletin, v. 45, pp. 1557-1571.

Lennon, G., 2000. Folly Beach, South Carolina: Tomorrow's Coastal Problems Today. *In:* A Compendium of Field Trips of South Carolina Geology with Emphasis on the Charleston, South Carolina Area. SC. Dept. of Natural Resources, Geological Survey, pp. 39-46.

McCants, C.Y., 1982. Evolution and Stratigraphy of a Sandy Tidal Flat Complex within a Mesotidal Estuary. M.S. Thesis, Dept. Geology, University of South Carolina, Columbia, SC, 97 pp.

Meade, R.H., 1969. Landward Transport of Bottom Sediments in Estuaries of the Atlantic Coastal Plain. Journal of Sedimentary Petrology, v. 39, pp. 222-234.

Monroe, J.S. and Wicander, R., 1998, Physical Geology. (3rd edition). Wadsworth Publishing Company, Belmont, CA, 646 pp.

Miller, P.P., 2006. Fripp Island: A History. History Press, Charleston, SC, 192 pp.

Mitsch, W.J. and Gosselink, J.G., 2000. Wetlands, 3rd edition. John Wiley and Sons, New York, NY, 920 pp.

Moslow, T.F. and Heron, D.S., 1978. Relict Inlets: Preservation and Occurrence in the Holocene Stratigraphy of Southern Core Banks, N.C. Journal of Sedimentary Petrology, v. 48, pp. 1275-1286.

Moslow, T.F., 1980. Stratigraphy of Mesotidal Barrier Islands. Ph.D. Thesis, Dept. Geology, University of South Carolina, Columbia, SC, 186 pp.

Murphy, C.H., 1995. Carolina Rocks! Sandlapper Publishing Co., Inc., Orangeburg, SC, 261 pp

Nelson, D.D., Harris, M.S., Wright, E., and Gayes, P.T., 2007. Coastal Plain History as Revealed by High Resolution Topographic Surveys, Horry County, SC (Abstract). Geological Society of America, Southeastern Section–56th Annual Meeting, Savannah, GA, 29–30 March 2007.

Nixon, Z., 2006. Wetland Community Change and Hydrologic Modification in the Santee River Delta. M.S. thesis, Duke University, Durham, NC, 74 pp.

O'Brien, M.P., 1931. Estuary Tidal Prisms Related to Entrance Areas. Journal of Civil Engineering, v. 1(8), pp. 738-793.

Oertel, G.F. and Larsen, M., 1976. Development Sequences in Georgia Coastal Dunes and Distribution of Dune Plants. Bulletin of Georgia Academy of Science, v. 34, pp. 35-48.

Pierce, J.W. and Colquhoun, D.J., 1970. Holocene Evolution of a Portion of the North Carolina Coast. Geological Society of America Bulletin, v. 81, pp. 3697-3714.

Pilkey, O.H., 2007. Beach Nourishment is Not the Answer. Business & Economic Review, v. 53, no. 2.

Pilkey, O.H., and Fraser, M.E., 2003. A Celebration of the World's Barrier Islands. Columbia University Press, New York, NY, 309 pp.

Pilkey, O.H. and Pilkey-Jarvis, L., 2006. Useless Arithmetic: Why Environmental Scientists Can't Predict the Future. Columbia University Press, New York, NY, 230 pp.

Pompe, J.J. and Rinehart, J.R., 2007. Beach Nourishment: A Taxing Question. Business & Economic Review, Jan-March issue.

Postma, H., 1967. Sediment Transport and Sedimentation in the Estuarine Environment. Estuaries, v. 83, pp. 158-79.

Prevost, C.K. and Wilder, E.L., 1972. Pawleys Island ...... a Living Legend. State Printing Company, Columbia, SC.

Pritchard, D.W., 1967. What is an Estuary: Physical Viewpoint. In: Lauff, G.W. (ed.), Estuaries, American Association for the Advancement of Science, Publication No. 83, pp. 3-5.

Rosen, P.S., 1975. Origin and Processes of Cuspate Spit Shorelines. In: Cronin, L.E. (ed) Estuarine Research 2. Academic Press, New York, NY, pp. 77-92.

Rubillo, T., 2006, Hurricane Destruction in South Carolina – Hell and High Water. History Press, Charleston, SC, 147 pp.

Ruby, C.H., 1981. Clastic Facies and Stratigraphy of a Rapidly Retreating Cuspate Foreland, Cape Romain, South Carolina. Ph.D. Thesis, Dept. of Geology, University of South Carolina, Columbia, SC, 207 pp.

Rust, B.R., 1978. Depositional Models for Braided Alluvium. In: Miall, A.D. (ed), Fluvial Sedimentology. Canadian Society of Petroleum Geologists, Memoir 5, pp. 605-625.

Rutledge, A., 1956. Santee Paradise: The Beautiful Wilderness Around Hampton Plantation. Bobbs-Merrill Co., Inc., Indianapolis, IN, 232 pp.

Rutledge, A.,1958. Home by the River. Bobbs-Merrill Co., Inc., Indianapolis, IN, 168 pp.

Rutledge, A., 1960, 1966 (revised edition). Deep River, The Complete Poems of Arichibald Rutledge. the R.L. Bryan Co., Columbia, SC, 635 pp.

Sexton, W.J., 1987. Morphology and Sediment Character of Mesotidal Shoreline Depostional Environments. Ph.D. Thesis, Dept. of Geology, University of South Carolina, Columbia, SC, 197 pp.

Sexton, W.J. and Hayes, M.O., 1982. Natural Bar-bypassing of Sand at a Tidal Inlet. Proc. 18th Coastal Engineering Conference, Capetown, South Africa, American Society of Civil Engineers, v. 2, pp. 1479-1495.

Sexton, W.J. and Hayes, M.O., 1991. The Geologic Impact of Hurricane *Hugo* and Post-storm Recovery Along the Undeveloped Coastline of South Carolina, Dewees Island to the Santee Delta. Journal of Coastal Research, Special Issue 8, pp. 275-290.

Sexton, W.J. and Hayes, M.O., 1996. Holocene Deposits of Reservoir-quality Sand Along the Central South Carolina Coastline. American Association of Petroleum Geologists Bulletin, v. 80, no. 6, pp. 831-855.

Sharitz, R.R., 1975. Forest Communities of Kiawah Island. *In:* Environmental Inventory of Kiawah Island, Environmental Research Center, Inc., Columbia, SC, pp. F-1 – F-43.

Shepard, F.P. and Wanless, H.R., 1971. Our Changing Coastlines. McGraw-Hill, New York, NY, 579 pp.

Silvester, R., 1977. The Role of Wave Reflection in Coastal Processes. Coastal Sediments '77, American Society of Civil Engineers, New York, NY, pp. 639-654.

Simons, D.B. and Richardson, E.V., 1962. Resistance to Flow in Alluvial Channels. American Society of Civil Engineers Trans., v. 127, pp. 927-953.

Staub, J.R., and Cohen, A.D., 1979. The Snuggedy Swamp of South Carolina: A Back-barrier Estuarine Coal-forming Environment. Journal of Sedimentary Petrology, v. 49, pp. 133-144.

Stalter, R., 1974. Vegetation in the Cooper River Estuary. Cooper River Environmental Study, S.C. Water Resources Commission, Columbia, SC.

Stapor, F.W. and Matthews, T.D., 1976. Mollusc C-14 ages in the Interpretation of South Carolina Barrier Island Deposits and Depositonal Histories. *In:* Hayes, M. O. and Kana, T. W. (eds), Terrigenous Clastic Depositional Environments. Technical Report 11-CRD, Geology Dept., University of South Carolina, Columbia, SC, pp. II-101 – II-114.

Stephen, M.F., Brown, P.J., FitzGerald, D.M., Hubbard, D.K., and Hayes, M.O., 1975. Beach Erosion Inventory of Charleston County, South Carolina: A Preliminary Report. South Carolina Sea Grant Technical Report 4, South Carolina Sea Grant Program, Charleston, SC, 79 pp.

Svetlichny, M., 1982. Short Term Impact of Beach Scraping and Backshore Restoration, Myrtle Beach, South Carolina. M.S. thesis, Dept. of Geology, University of South Carolina., Columbia,SC, 107 pp.

Sydow, J. and Roberts, H.H., 1994. Stratigraphic framework of a Late Pleistocene Shelf-edge Delta, Northeast Gulf of Mexico. American Association of Petroleum Geologists Bulletin, v. 78, no. 8, pp. 1276-1312.

Swift, D.J.P., 1975. Barrier-island Genesis: Evidence from the Central Atlantic Shelf, Eastern USA. Sedimentary Geology, v.14, pp. 1-43.

Talwani, P., 1982. Internally Consistent Pattern of Seismicity near Charleston, South Carolina. Geology, v. 10, pp. 654-658.

Tannehill, I.R., 1956. Hurricanes. Princeton University Press, Princeton, NJ, 308 pp.

Taylor, J.R., Cardamone, M.A., and Mitsch, W.J., 1990. Bottomland Hardwood Forests: Their functions and Values. *In:* Gosselink, J.G., Lee, L.C., ad Mitsch, W.J. (eds), Ecological Processes and Cumulative Impacts: Illustrated by Bottomland Hardwood Wetland Ecosystems. Lewis Publishing Co., Chelsea, MS, pp. 13-86.

Teal, J.M., 1958. Distribution of Fiddler Crabs in Georgia Salt Marshes. Ecology, v. 39, pp. 185-193.

Tye, R.S., 1981. Geomorphic Evolution and Stratigraphic Framework of Price and Capers inlets, South Carolina. M.S. Thesis, Dept. of Geology, University of South Carolina, Columbia, SC, 144 pp.

Tye, R.S., 1984. Geomorphic Evolution and Stratigraphy of Price and Capers Inlets, South Carolina. Sedimentology, v. 31, pp. 655-674.

U.S. Army Corps of Engineers, 1971. National Shoreline Study, Washington, DC

U.S. Army Corps of Engineers, 2002. Coastal Engineering Manual. Engineer Manual 1110-2-1100, U.S. Army Corps of Engineers, Washington, DC.

Van Straaten, L.M.J.U., 1950. Environment of Formation and Facies of the Wadden Sea Sediments. Koninkl. Ned. Aardrijkskde Genoot., v. 67, pp. 94-108.

Vaughan, C., 1975. Pawleys-As It Was. Hammock Shop, Pawleys Island, SC, 73 pp.

Ward, L.G. and Domeracki, D.D., 1978. The Stratigraphic Significance of Back-barrier Tidal Channel Migration (Abstract). Proceedings of the Geological Society of America Meeting (SE Sec.), v.10, p. 20.

Wentworth, C.K., 1922. A Scale of Grade and Class Terms for Clastic Sediments. Journal of Geology, v. 30(5), pp. 377-392.

Wiedemann, H. U., 1972. Application of Red-lead to the Detection of Dissolved Sulfide in Waterlogged Soils. Zeitschr. Pflanzenernahrung Bodenkunde, v. 133, pp. 73-81.

Winker, A.D. and Howard, J.D., 1977. Correlation of Tectonically Deformed Shorelines of the Southern Atlantic Coastal Plain. Geology, v. 5, pp. 123-127.

Wood, F.J., 1982. Tides. *In:* Schwartz, M.L. (ed.), The Encyclopedia of Beaches and Coastal Environments. Hutchinson Ross Pub. Co., Stroudsburg, PA, pp. 826-837.

Wright, E. and Forman, S., 2007, Geologic Studies of Carolina Bays in Northeastern South Carolina (Abstract). Southeastern Section, Geological Society of America–56th Annual Meeting, Savannah, GA, 29–30 March 2007.

Zabawa, C.F., 1976. Estuarine Sediments and Sedimentation Processes in Winyah Bay South Carolina. *In:* Hayes, M.O. and Kana, T.W. (eds), Terrigenous Clastic Depositional Environments. Technical Report 11-CRD, Geology Dept., University of South Carolina, Columbia, SC, pp. II-64 - II-78.

Zarillo, G.A., Ward, L.G., and Hayes, M.O., 1985. An Illustrated History of Tidal Inlet Changes in South Carolina. South Carolina Sea Grant Consortium, Charleston, SC, 76 pp.

Zenkovitch, V.P., 1967. Processes of Coastal Development. Interscience, New York, NY, 738 pp.

INDEX

A

ACE Basin 225–226
alligators 241
amphidromic systems 25

B

barrier islands 2, 44
  drumstick-shaped 46
  origin 46, 51
  prograding 44
  transgressive 44
beach cusps 221
beach erosion 81
  beach nourishment 93, 138
  causes 82
  deficits in sand supply 82
  prevention 91
  sea-level rise 87
Beachwalker Park 209
Bear Island WMA 232
black-water rivers 83, 123
Blue Ridge Province 16
Boone Hall Plantation 191
bottomland hardwoods 125
Brookgreen Gardens 150
brown pelicans 149
Bull Island area 179
Bulls Bay 176

C

Cape Romain 170
Cape Romain National Wildlife Refuge 185
Captain Sams Inlet 213

Carolina Bays 145–149
Coastal Discovery Museum 251
coastal dunes 76
Coastal Plain 7, 12
coastal plain rivers 83, 123
Congaree National Park 262
cuspate forelands 171
cut bank 117–119
Cypress Gardens 261
cypress-tupelo swamps 125

D

Daufuskie Island 253
deltas 61, 83
depositional coasts 21, 22, 23, 50
Donnelly WMA 233
downdrift offset 46, 50
Drayton Hall 259
drift 46

E

earthquakes 42
ebb-tidal delta 46, 53, 56–57
Edisto Beach State Park 217
Ernest F. Hollings ACE Basin NWR 231
estuaries
  definition 56
  general characteristics 58–59
  in South Carolina 61
  zone of the turbidity maximum 59

F

fall line 3, 13, 123, 125
flood-tidal delta 53

Folly Beach 191
Folly Beach County Park 197
foredunes 45, 47, 63
Francis Beidler Forest 261
Francis Marion National Forest 260
Fripp Island 245

G

geologic time 9
geomorphology 1
Georgia Bight 11, 28, 43
ghost crab 210–211
ghost shrimp 210, 212
Gondwana 7
Grand Strand 135
Great Swamp Sanctuary 230
Green Shell Enclosure Heritage Preserve 252
Grove Plantation 231

H

hammock 257
Hampton Plantation State Historic Site 168
heavy minerals 72, 74–75
highstand 14
Hilton Head Island 248
Holocene Epoch 15
hooded mergansers 189
Hopsewee Plantation 167
horseshoe crabs 108–111
Hunting Island State Park 238
Huntington Beach State Park 149
hurricane 39
hurricane Hugo 94

I

Ice Ages 13
igneous rocks 8
Isle of Palms 186

K

Kiawah Island 200

L

Lewis Ocean Bay Heritage Preserve 145–146
Lighthouse Inlet 197
loblolly pines 263
longshore current 33, 35
Low Country 225
lowstand 13, 58
lowstand delta 15, 129
lowstand valleys 14, 115

M

macrotidal 21
Magnolia Gardens 260
marshes
 brackish 106
 freshwater 104
 salt 106
meandering channels 114, 118
mesotidal 21, 22, 53
metamorphic rocks 9
microtidal 21, 22
Middleton Place 260
mixed-energy coasts 22
Morris Island 191
Myrtle Beach State Park 142

N

neap tides 25
northern gannet 149
North Inlet 158

P

painted bunting 209–210
Pangaea 2, 7, 10
parks
 Beachwalker Park 209
 Congaree National Park 262
 Edisto Beach State Park 217
 Folly Beach County Park 197
 Huntington Beach State Park 149
 Myrtle Beach State Park 142
 Woods Bay State Park 149
peat formation 228
Pickett Recreation Area 188
piedmont 17
piedmont rivers 123
Pinckney Island National Wildlife Refuge 251
plate tectonics 19, 20
Pleistocene Epoch 13, 14, 44, 58, 87, 147, 235, 236
point bars 118
Port Royal Sound 245

R

red-cockaded woodpecker 260
red-shouldered hawk 262–263
rhythmic topography 33
rift basins 9

S

salt pruning 142

sand beaches 63, 66
  antidune 70, 73
  beach cycle 67
  beach profile 64–66
  physical sedimentary features 68–69
  zone of dynamic change 63–64
Santee/Pee Dee Delta Region 155
Savannah National Wildlife Refuge 255
Savannah River Delta 254
Seabrook Island 213
sea-level rise 87, 89–90, 265
seas 28–29
sedimentary rocks 9
sediment types 23
Snuggedy Swamp 228–229
spring tides 25, 26
St. Helena Island 239
St. Helena Sound 225, 227–228
storm surge 39–40
Sullivans Island 186
swell 28–30

T

Tertiary Period 12
the Georgia Bight 11, 28, 43
tidal channels 114
tidal flats 108
  dominant biogenic features 113
  sandy 108
  sheltered muddy 109
tidal inlets 50
  general model 54
tidal rice fields 159
Tom Yawkey Wildlife Center Heritage Preserve
     165–166

U

upper bottomland hardwoods 125

V

Victoria Bluff Heritage Preserve 250

W

Waddell Mariculture Center 250
Wambaw Swamp Wilderness 261
wave diffraction 33
wave refraction 34
waves 27–36
  plunging 31
  spilling 31
  surging 31

Winyah Bay 155–156
Woods Bay State Park 149
wrack 76

ABOUT THE AUTHORS

Miles O. Hayes

Dr. Miles O. Hayes is a coastal geomorphologist with over 50 years of research experience. He has authored over 250 articles and reports and three books on numerous topics relating to tidal hydraulics, river morphology and processes, beach erosion, barrier island morphology, oil pollution, and petroleum exploration. Hayes' teaching experience includes both undergraduate and graduate courses while a Professor at the Universities of Massachusetts and South Carolina. Seventy-two graduate students received their degrees under his supervision, most of whom are now leaders in their respective academic, government and industry positions. He is considered to be the "Father of Coastal Geology." A review of his 1999 book – Black Tides, said "A skilled raconteur, Hayes tells engrossing stories of responding to most of the recent, headline-grabbing oil spills, including the Gulf War spills, the *Exxon Valdez*, the *Amoco Cadiz* spill in France, and the *Ixtoc I* blowout in Mexico. Interspersed among them are personal events and adventures, such as his survival of a plane crash while mapping a remote part of Alaska." He is Chairman of the Board of Research Planning, Inc. (RPI), a science technology company located in Columbia, S. C.

Jacqueline Michel

Dr. Jacqueline Michel is an internationally recognized expert in oil and hazardous materials spill response and assessment with a primary focus in the areas of oil fates and effects, non-floating oils, shoreline cleanup, alternative response technologies, and natural resource damage assessment. As of this time, she has participated in research projects in 33 countries. Much of her expertise is derived from her role, since 1982, as part of the Scientific Support Team to the U.S. Coast Guard provided by the National Oceanic and Atmospheric Administration. Under this role, she is on 24-hour call and provides technical support for an average of 50 spill events per year. She leads shoreline assessment teams and assists in selecting cleanup methods to minimize the environmental impacts of the spill. She has written over 150 manuals, reports, and scientific papers on coastal resource impacts, mapping, and protection. As a member of the Ocean Studies Board at the National Academy of Sciences for four years, she served on four National Research Council committees (chairing two), and is a Lifetime Associate of the National Academies. One of the original founders of RPI, which started in 1977, she now serves as the company President.